Green Energy and Technology

Climate change, environmental impact and the limited natural resources urge scientific research and novel technical solutions. The monograph series Green Energy and Technology serves as a publishing platform for scientific and technological approaches to "green"—i.e. environmentally friendly and sustainable—technologies. While a focus lies on energy and power supply, it also covers "green" solutions in industrial engineering and engineering design. Green Energy and Technology addresses researchers, advanced students, technical consultants as well as decision makers in industries and politics. Hence, the level of presentation spans from instructional to highly technical.

Indexed in Scopus.

Indexed in Ei Compendex.

Nakorn Tippayawong · Thossaporn Onsree ·
James Moran

Machine Learning Applications for Bioenergy Conversion

Springer

Nakorn Tippayawong
Department of Mechanical Engineering
Chiang Mai University
Chiang Mai, Thailand

Thossaporn Onsree
Molinaroli College of Engineering
and Computing
University of South Carolina
Columbia, SC, USA

James Moran
Department of Mechanical Engineering
Chiang Mai University
Chiang Mai, Thailand

ISSN 1865-3529 ISSN 1865-3537 (electronic)
Green Energy and Technology
ISBN 978-981-96-8085-6 ISBN 978-981-96-8086-3 (eBook)
https://doi.org/10.1007/978-981-96-8086-3

This Springer imprint is published by the registered company Springer Nature Singapore Pte Ltd.
The registered company address is: 152 Beach Road, #21-01/04 Gateway East, Singapore 189721, Singapore

To help in our own small way reach a renewable future. And to those quietly working to understand nature a little better, and to those trying, in small ways, to make the future a bit cleaner.

Preface

The global shift toward sustainable energy has never been more critical. As the effects of climate change and the limitations of fossil fuels become increasingly evident, the search for renewable energy solutions has intensified. Among these solutions, bioenergy—a resource derived from organic materials such as plants, waste, and algae—offers tremendous promise. Yet, despite its potential, the field of bioenergy faces significant challenges in optimizing processes, improving efficiency, and addressing variability in raw materials. This is where machine learning (ML), a powerful branch of artificial intelligence, steps in as a game-changing tool.

This book, *Machine Learning Applications for Bioenergy Conversion*, is written to bridge the gap between two dynamic fields: bioenergy and machine learning. It aims to provide readers with an accessible yet comprehensive guide to understanding how ML can unlock the full potential of bioenergy systems. Unlike technical journal articles that cater to specialized audiences, this book is designed to be approachable for a broad spectrum of readers, including researchers and industry professionals interested in sustainable energy solutions. It is not designed for beginners or people with no experience in renewable energy who seek an introductory book. A background in renewable energy and its basic terminology is assumed.

The chapters within this book explore how ML techniques are applied to various bioenergy conversion processes, including hydrothermal carbonization, co-pyrolysis, biodiesel production, anaerobic digestion, and hydrothermal processes. Through practical examples, case studies, and a focus on user-friendly language, we unravel the complex interplay between ML algorithms and bioenergy technologies. Each chapter highlights specific challenges and demonstrates how ML models—such as Random Forest, Support Vector Machines, and Extreme Gradient Boosting—can predict outcomes, optimize processes, and provide insights to enhance energy yields and sustainability.

This book reflects the culmination of diverse research efforts and practical applications from experts in the field. It is our hope that readers will not only gain a deeper understanding of these cutting-edge technologies but also feel inspired to contribute to the ongoing development of sustainable energy systems.

Chiang Mai, Thailand — Nakorn Tippayawong
Columbia, USA — Thossaporn Onsree
Chiang Mai, Thailand — James Moran

Acknowledgments This work was partially supported by the National Research Council of Thailand (contract no. N42A671047) and Chiang Mai University.

Competing Interests The authors have no competing interests to declare that are relevant to the content of this manuscript.

About This Book

This book explores the application of machine learning (ML) techniques to optimize bioenergy conversion processes. It addresses the growing global demand for renewable energy sources by focusing on the use of biomass, a promising alternative to fossil fuels. The text delves into various ML algorithms and their ability to predict and enhance bioenergy outcomes, such as biocrude yields, energy content, and the efficiency of processes like biodiesel production, anaerobic digestion, and hydrothermal processes. Each chapter focuses on a specific bioenergy process, illustrating how machine learning techniques can enhance various stages of these processes. Special emphasis is placed on advanced methods for handling complex feedstocks, such as wet biomass, and predicting key factors like yield and energy quality. It also covers how machine learning aids in understanding the interactions between biological, chemical, and processing variables, ultimately leading to more efficient and sustainable bioenergy production. The book combines fundamental scientific insights with practical applications, providing both theoretical knowledge and actionable guidelines for optimizing biomass conversion technologies.

By integrating theoretical insights with practical applications, the book aims to bridge the gap between cutting-edge AI methodologies and bioenergy technologies. This book is primarily aimed at advanced readers, including researchers, engineers, and practitioners in the fields of bioenergy, renewable energy, and artificial intelligence, who are looking to deepen their understanding of the sophisticated techniques and methodologies involved in these interdisciplinary areas. It also serves as a valuable resource for graduate students with a strong foundation in these subjects, who seek to explore the complex applications of machine learning in the development of sustainable energy solutions. By offering in-depth discussions and advanced analytical tools, this book provides a comprehensive guide for those looking to apply cutting-edge machine learning strategies to real-world bioenergy challenges, while also fostering a deeper understanding of the scientific principles at the intersection of energy production and technology.

Contents

About the Authors

Dr. Nakorn Tippayawong received B.E. in Mechanical Engineering and Ph.D. in Internal Combustion Engines from Imperial College London, UK. He is a professor at the Department of Mechanical Engineering, Chiang Mai University. His research interests include biomass conversion, energy efficiency improvement, decarbonization technology, and emission control.

Dr. Thossaporn Onsree holds a Ph.D. in Mechanical Engineering from Chiang Mai University, Thailand. He is currently a post-doctoral fellow in the Molinaroli College of Engineering and Computing, University of South Carolina, USA. His research interests include catalytic and non-catalytic biomass conversions, hydrogen carriers, high-throughput experimentation, and machine learning and artificial intelligence applications. He applies machine learning to explore insight into systems of biomass conversions, which will help improve and optimize practical/industry processes. With high-throughput experimentation and artificial intelligence, he is developing a tool to accelerate the discovery of novel materials that can carry/transport hydrogen effectively, especially for long-distance transportation.

Dr. James Moran received a Masters degree and Ph.D. degree in Mechanical Engineering from the Massachusetts Institute of Technology. He is currently an Associate Professor at the Department of Mechanical Engineering in Chiang Mai University, Thailand, teaching in energy related fields. His research interests include biogas upgrading, biomethane grids, biomethane low-pressure storage, meso-scale combustion, aerosol generation, and pollution modeling.

All the authors have published numerous peer-reviewed papers on biomass, contributing extensively to the body of knowledge in this field. This book represents a culmination and synthesis of their collective work, bringing together their individual research and findings into a comprehensive, cohesive resource. It reflects their expertise and years of experience in studying biomass, covering a wide range of topics from production techniques to applications, while also offering fresh insights and innovative perspectives. By compiling these efforts into one volume, the authors aim to provide a thorough and accessible guide that builds on the latest advancements in biomass research and practice.

List of Figures

List of Tables

Chapter 1
An Introduction to Machine Learning in Bioenergy

Abstract Machine Learning (ML) and its transformative role in advancing bioenergy technologies. As the global need for sustainable energy solutions intensifies, bioenergy stands out as a renewable resource with immense potential. However, its development is often hindered by the complexity and variability of biomass materials and conversion processes. Machine learning addresses these challenges by leveraging large datasets to predict outcomes, optimize operating conditions, and uncover hidden patterns. This chapter provides an overview of how ML techniques, such as Random Forest, Neural Networks, and Extreme Gradient Boosting, are applied to enhance bioenergy processes like hydrothermal carbonization, pyrolysis, and biogas reforming. It sets the stage for the following chapters, which explore detailed applications and case studies, illustrating how ML drives innovation and efficiency in the quest for sustainable energy.

Keywords Bioenergy · Machine learning · Algorithm selection · Process optimization

1.1 Introduction

Energy is a fundamental part of our lives. From powering our homes to fueling industries, we rely on energy to drive progress and innovation. However, the growing demand for energy and the reliance on fossil fuels have led to environmental challenges such as global warming and resource depletion. This has prompted scientists and engineers to explore alternative energy sources, one of which is bioenergy—energy derived from biological materials such as plants, agricultural waste, and organic residues.

In recent years, the application of machine learning (ML) has emerged as a powerful tool in advancing bioenergy research and technology. Machine learning, a subset of artificial intelligence (AI), enables computers to analyze vast amounts of data, identify patterns, and make predictions without explicit programming. This chapter introduces the concept of machine learning, its potential in bioenergy applications, and how it is reshaping the field of renewable energy.

N. Tippayawong et al., *Machine Learning Applications for Bioenergy Conversion*, Green Energy and Technology, https://doi.org/10.1007/978-981-96-8086-3_1

1.2 What is Machine Learning?

Machine learning is a field of study that focuses on teaching computers to learn from data. Instead of following fixed instructions, machine learning models use algorithms to analyze information and improve their performance as they process more data. Think of it as training a computer to recognize patterns, such as distinguishing between different types of plants based on images or predicting the energy output of a biofuel based on its chemical properties. Machine learning is a branch of artificial intelligence (AI) that enables computers to learn from data and make predictions or decisions without being explicitly programmed. It involves training models on large datasets, allowing them to recognize patterns, identify relationships, and solve problems. These models can predict outcomes, optimize processes, and provide valuable insights that are often difficult for humans to uncover on their own.

There are three main types of machine learning:

1. Supervised Learning: The model is trained on a labeled dataset, meaning each input has a corresponding output. For example, predicting the energy content of biomass based on its composition.
2. Unsupervised Learning: The model identifies patterns in data without labeled outputs. This can be used to group similar types of biomass or identify new trends in bioenergy production.
3. Reinforcement Learning: The model learns by interacting with its environment and receiving feedback. This method can optimize processes, such as improving the efficiency of a bioenergy plant.

1.3 Why Machine Learning in Bioenergy?

The growing demand for sustainable energy has sparked a global interest in bioenergy, a renewable energy source derived from organic materials such as plants, agricultural waste, and even algae. As the world seeks to combat climate change and reduce dependence on fossil fuels, bioenergy has emerged as a promising solution. However, unlocking the full potential of bioenergy requires tackling complex challenges, from optimizing production processes to improving energy yields. This is where machine learning (ML) plays a transformative role. Bioenergy production involves complex processes, from converting raw biomass into usable energy to optimizing the performance of renewable energy systems. These processes generate massive amounts of data, and understanding this data is key to improving efficiency and sustainability. Machine learning can process and analyze this data far more efficiently than traditional methods, providing valuable insights to:

- Predict the yield and quality of biofuels based on raw material properties.
- Optimize conversion processes such as hydrothermal carbonization, pyrolysis, and biodiesel production.

- Reduce waste and emissions by identifying inefficiencies in production.
- Model and predict the environmental impact of bioenergy systems.

1.3.1 Real-World Applications of Machine Learning in Bioenergy

In bioenergy, machine learning serves as a powerful tool to address various challenges. For example, it can predict how different types of biomass will perform in conversion processes like hydrothermal carbonization (HTC), pyrolysis, and anaerobic digestion. ML can also help optimize operating conditions, such as temperature, pressure, and reaction time, to maximize energy yields while minimizing waste and environmental impacts. By analyzing large datasets, machine learning models can identify the best feedstock combinations, improve fuel properties, and even reduce greenhouse gas emissions during production. The integration of machine learning into bioenergy research has already shown promising results. For instance:

- Biomass Conversion: Machine learning algorithms like Extreme Gradient Boosting (XGB) and Random Forest (RF) are used to predict the energy content, yield, and quality of biofuels produced from biomass.
- Process Optimization: Models can optimize parameters such as temperature, pressure, and reaction time in processes like hydrothermal carbonization, leading to higher efficiency and lower costs.
- Environmental Impact Assessment: By analyzing data from bioenergy systems, machine learning can predict carbon emissions, helping researchers develop more sustainable practices.
- Material Characterization: Machine learning helps identify the best types of biomass for energy production based on their chemical and physical properties.

1.3.2 Bridging the Gap Between Data and Action

The adoption of machine learning in bioenergy is not just about improving existing processes; it's about unlocking new possibilities. With the ability to handle complex datasets and uncover hidden relationships, machine learning offers a pathway to innovations that were previously out of reach. For example, machine learning can identify how subtle changes in biomass composition affect energy output or predict the long-term sustainability of bioenergy systems under different scenarios.

One of the key advantages of machine learning is its ability to handle the complexity and variability of bioenergy systems. Biomass is a diverse and inconsistent resource, with properties that can vary significantly depending on the source and processing method. Traditional methods of analysis often struggle to capture these variations, but ML excels at managing complex, high-dimensional datasets. It can uncover hidden relationships between variables, enabling more accurate predictions and better decision-making.

1.4 The Structure of this Book

In this textbook, the intersection of machine learning and bioenergy conversion processes is explored. The chapters delve into specific areas where ML has been applied successfully, such as hydrothermal carbonization, co-pyrolysis, biodiesel production, and biogas reforming. How different machine learning algorithms, including Random Forest, Neural Networks, and Extreme Gradient Boosting (XGB), are used to model and optimize these processes are examined. Each chapter highlights real-world case studies and research findings, demonstrating the practical benefits of combining ML with bioenergy technologies. The following is a brief summary of the contents of the remaining Chapters.

Chapter 2 provides an overview of artificial intelligence (AI) and machine learning (ML), emphasizing their transformative impact across industries, particularly bioenergy. It introduces key machine learning algorithms (MLAs) such as random forest (RF), support vector machines (SVM), gradient boosting (GB), and kernel ridge regression (KRR), which are foundational to the book's analyses. ML is highlighted for its ability to model complex biomass conversion processes, predict biocrude oil yields, and estimate calorific values of torrefied biomass using large datasets. Regression techniques, including linear and multiple regression, are explained as tools for predicting relationships between bioenergy variables like temperature and biomass yield. Decision trees are detailed as supervised learning models that simplify complex decision-making through hierarchical structures, with examples in bioenergy crop suitability and coal combustion decisions. Challenges like overfitting, sensitivity to data changes, and limited generalization are noted as limitations of decision trees. Overall, the chapter underscores ML's role in advancing sustainable bioenergy systems through predictive modeling and data-driven insights.

Chapter 3 offers a comprehensive review of the use of ionic liquid solvents (ILSs) for biomass pretreatment, focusing on their role in generating cellulose-rich materials (CRMs) and the application of machine learning to optimize this process. The authors detail how ILSs, particularly when combined with co-solvents, can effectively overcome the structural recalcitrance of lignocellulosic biomass by selectively dissolving lignin and hemicellulose, thereby enriching the cellulose content. What sets this work apart is its integration of machine learning algorithms to predict key CRM properties—specifically, the cellulose enrichment factor and solid recovery—based on a diverse dataset of biomass and process parameters. The random forest model emerged as the most effective tool, demonstrating high predictive accuracy and emphasizing the dominant influence of biomass type and ILS operating conditions. The chapter not only reinforces the potential of ILSs in advancing sustainable bio-resource processing but also highlights the value of data-driven modeling in improving efficiency and guiding future solvent and process design.

Chapter 4 presents a well-structured and insightful application of machine learning to predict the yield of torrefied biomass, offering a valuable advancement over traditional modeling approaches. By leveraging a comprehensive dataset of over 800 observations and fourteen features spanning both biomass properties and torrefaction conditions, the study moves beyond simplistic models to uncover complex, non-linear relationships. The gradient tree boosting algorithm demonstrated strong predictive performance, with high accuracy and low error, reinforcing its suitability for modeling such multidimensional data. Importantly, the chapter identifies six key features influencing solid yield, notably the roles of volatile matter and process temperature, and highlights how machine learning can clarify their individual and combined effects. The thoughtful preprocessing of data, including normalization via Yeo-Johnson transformation, enhances model robustness. Overall, the work not only demonstrates the technical potential of machine learning in biomass research but also provides practical insights for optimizing torrefaction, reinforcing the value of data-driven strategies in renewable energy development.

Chapter 5 investigates the co-pyrolysis of biomass and plastic waste to enhance bio-oil yield (BOY) and reduce environmental burdens, highlighting the role of synergistic effects (SE) between materials. Traditional modeling techniques struggle with the complexity of multiple interacting variables, so the study adopts machine learning—specifically, the XGBoost algorithm paired with Shapley Additive Explanations (SHAP)—to create accurate and interpretable models using 26 input features. The dataset, curated from 360 experimental cases, includes biomass composition, plastic types, and pyrolysis conditions, and was balanced using the SMOGN technique to address underrepresented outcomes. The models achieved high prediction accuracy, with SHAP analysis revealing that the plastic-to-biomass ratio, feedstock properties, and pyrolysis temperature were key predictors. Importantly, the analysis uncovered interactions—such as the catalytic role of biomass ash—that traditional models often miss. This machine learning framework, supported by advanced data balancing and interpretability tools, offers a powerful approach to understanding and optimizing co-pyrolysis systems, promoting the production of higher-quality, energy-dense bio-oils while supporting circular economy principles.

Chapter 6 explores how machine learning (ML), particularly extreme gradient boosting (XGB), can accurately predict biodiesel yields from transesterification processes involving various catalysts—homogeneous, heterogeneous, and enzyme-based. Using a well-curated dataset of over 1,000 data points collected from extensive literature, the study identifies key operational and feedstock factors influencing biodiesel production, such as alcohol-to-oil ratio, reaction time, temperature, catalyst concentration, and fatty acid profiles. For instance, linoleic acid proved most influential with homogeneous catalysts, behenic acid with heterogeneous ones, and reaction time for enzymatic processes. The model achieved high predictive accuracy using techniques like tenfold cross-validation and feature importance analysis. The fatty acid composition—saturation, chain length, and free fatty acid content—emerged as critical to biodiesel yield and quality, impacting parameters like cold flow properties and oxidation stability. The chapter also details the use of artificial neural networks and response surface methodology in biodiesel research, underscoring the value of ML in optimizing production and informing catalyst selection. Overall, the integration of ML tools enhances the understanding of complex interactions in biodiesel synthesis and supports the development of more efficient, sustainable biofuel technologies.

Chapter 7 demonstrates the effective application of machine learning (ML) techniques to predict and optimize specific methane yields (SMY) from lignocellulosic biomass (LB) in anaerobic digestion (AD) systems. Using a dataset of 205 entries with 14 input features covering biomass properties and operating conditions, random forest (RF) emerged as the most accurate model. Cellulose content was identified as the most influential factor, and co-digestion with manure—particularly at a 1:1 ratio under standard organic loading rates—was found to enhance methane yields, confirmed experimentally with SMY reaching 79.2% of the RF-predicted value. The study used partial dependence plots to analyze feature impacts and interactions, optimized the biomass-to-manure ratio across various loading rates, and validated the model with pilot-scale experiments using Giant Juncao Grass, rice straw, and corn stover. Results underscore ML's potential to improve biogas production efficiency by integrating both substrate characteristics and operational parameters in model development, and by providing practical recommendations for reactor operation.

Chapter 8 explores hydrothermal carbonization (HTC) as an environmentally friendly method for converting wet biomass into hydrochar, a carbon-rich solid useful as biofuel or soil amendment. HTC operates under high temperature and pressure without requiring biomass drying and involves complex chemical reactions such as hydrolysis, dehydration, decarboxylation, aromatization, and recondensation. To simplify analysis and enhance understanding, the chapter applies Extreme Gradient Boosting (XGB), a powerful machine learning algorithm, to predict hydrochar yield, energy content, ash content, and elemental composition (carbon, hydrogen, oxygen) based on a dataset of 1,070 entries with 16 input features. Compared to prior studies with smaller datasets and narrower predictive goals, this work demonstrates improved accuracy using modern techniques like tenfold cross-validation, hyperparameter tuning, and explainability tools such as Shapley values and SHAP plots. The findings reveal strong correlations between predicted and actual outcomes, providing actionable insights into the key factors governing HTC and highlighting the utility of machine learning for modeling biomass conversion processes.

Chapter 9 investigates how machine learning (ML) can accurately predict both biocrude yield (BY) and energy content (higher heating value, HHV) from hydrothermal liquefaction (HTL) of wet biomass and waste materials. Drawing on a dataset of 325 experimental observations across 17 input features—including biomass composition and processing conditions—the study evaluates four advanced ML algorithms (SVR, KRR, RF, and XGB), using rigorous tenfold cross-validation and optimized hyperparameters. Among these, XGB consistently delivers the highest accuracy, with low normalized errors. The analysis reveals that process temperature is the single most influential variable, while feedstock characteristics collectively account for over half of the predictive power. To deepen understanding of factor interactions, the study uses partial dependence plots (PDPs) for both single- and two-variable relationships. Compared to prior HTL modeling efforts, this work is novel in its simultaneous prediction of yield and energy content, systematic feature importance analysis, and use of a robust modeling framework. Overall, the findings highlight the potential of ML—particularly XGB—as a powerful tool for optimizing biofuel production from complex and variable biomass feedstocks.

By the end of this textbook, readers will gain a comprehensive understanding of how machine learning can revolutionize bioenergy, driving innovation and sustainability in this critical field. Whether you are a student, researcher, or professional, this knowledge will equip you to contribute to a greener, more energy-efficient future. Together, let's uncover how data and algorithms are shaping the next era of renewable energy.

Chapter 2
Machine Learning Algorithms for Bioenergy

Abstract Artificial Intelligence (AI) is transforming how tasks are performed by machines, particularly through computer systems capable of calculation, learning, and responding to stimuli in ways that mimic human cognition. A critical subset of AI is Machine Learning (ML), which focuses on the learning process. Much like the human brain, ML systems analyze data inputs, retain relevant information, and produce outputs as numerical or coded representations for decision-making or display purposes. The potential of AI and ML to revolutionize industries has made them central topics of discussion in recent years. This chapter introduces the types of machine learning algorithms (MLA) that are used throughout this book. It will serve as a reference source as various MLA get referred to in different chapters but the description of how they operate is discussed here. They are four main MLA used:

1. Random forest (RF)
2. Support vector machines (SVM)
3. Gradient boosting (GB)
4. Kernel ridge regression (KRR).

There are other algorithms, lasso, normal ridge regression, decision trees, adaptive boosting, k-nearest neighbors, that were evaluated in specific problem areas. They are not discussed in this chapter but introduced instead where they appear in this manuscript. In addition before any algorithm can operate, datasets need preprocessing. This mathematical operation is discussed here. The methods of interpreting results and evaluating their quality are also discussed here.

Keywords Machine learning algorithms · Bioenergy · Random forest · Kernal ridge regression · Gradient boosting · Support vector machines

2.1 Machine Learning Algorithms

The development of ML has enabled it to be integrated into various aspects of everyday life, including but not limited to medicine, marketing, sports, mass media, and both heavy and light industries. The rapid progress in this field has resulted in the

N. Tippayawong et al., *Machine Learning Applications for Bioenergy Conversion*, Green Energy and Technology, https://doi.org/10.1007/978-981-96-8086-3_2

creation of numerous technological applications that are currently in use (Jentzer et al. 2023). One prominent application of ML in the field of energy production is its use in the prediction of biocrude oil from the hydrothermal liquefaction process of biomass with accuracy rates of up to 90% (Katongtung et al. 2022). Another example is the utilization of ML to predict the value of cellulose-rich materials obtained from biomass pretreatment with ionic liquid solvents (Phromphithak et al. 2021). ML has also been employed in the prediction of yields of solid products from biomass torrefaction (Onsree and Tippayawong 2021).

Complex processes are required to convert raw biomass into bioenergy. Modeling their performance is difficult, and thus knowing the yields a priori is challenging. Models are based on either the reaction kinetics or empirical prediction models, such as simple regression and response surface methodology. They are validated using experimental data but the specific nature of the data limits the reliability and generalization of the models. To assess multidimensional relationships in biomass conversion processes, machine learning (ML) techniques can be applied, using bigger datasets. ML is a sub-branch of artificial intelligence that can automatically learn and recognize patterns from a dataset without being explicitly programmed. ML can access any kinds of data and extract knowledge outside of the original dataset. For example, Zhu et al. (2019) used properties from raw biomass to make predictions of solid yield and carbon content from biomass pyrolysis using a random forest ML algorithm. Based on a 12-feature pyrolysis dataset, they found that raw biomass features, such as ash content, hemicellulose content, and nitrogen content, had a 35% contribution on the solid yield. Garcia-Nieto et al. (2019) applied ML to estimate calorific values of the torrefied biomass as a function of raw biomass feedstock characteristics in combination with torrefaction conditions. Raw biomass properties, such as volatile matter, fixed carbon, and atomic O/C ratio, were found to have a high influence on calorific values of torrefied biomass.

2.1.1 Decision Trees

A decision tree is a type of supervised machine learning used to categorize or make predictions based on how a previous set of questions were answered. The model is trained and tested on a set of data that contains the desired categorization. Before discussing how a decision tree works, let's define some terminology. Figure 2.1 can be used as a visual aid.

- Root node: The root node is the base of the decision tree, which represents the starting point for the decision-making process. It is the topmost node in the tree and is used to split the data into two or more subsets based on a specific condition. The root node is typically chosen based on the most informative feature in the dataset
- Splitting: Splitting is the process of dividing a node into multiple sub-nodes. This is done by selecting a feature and a threshold value, and then splitting the data into

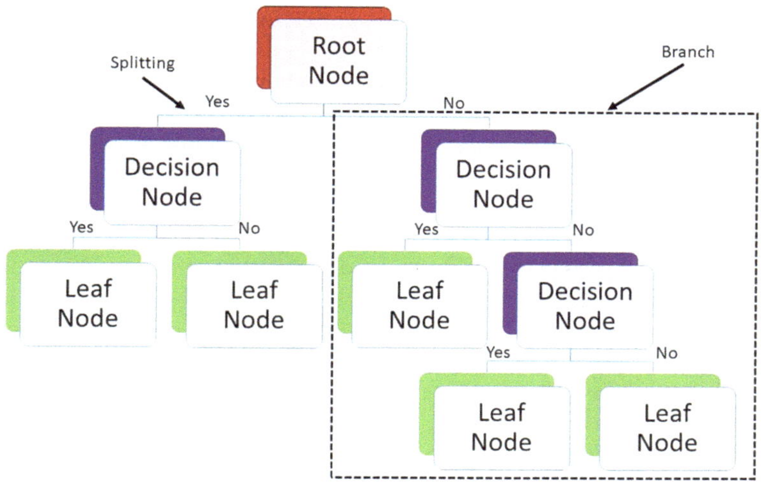

Fig. 2.1 Terminology used for decision trees

two subsets based on whether the feature value is above or below the threshold. Splitting is repeated recursively for each subset until a stopping criterion is met. An example is given as follows:

- Select a Feature and Threshold: Imagine a dataset with many variables (features) like temperature, soil quality, or sunlight hours. One of these features is selected along with a specific threshold value
- Divide the Data: The dataset is split into two groups based on whether the feature value is above or below the chosen threshold. For example, if the feature is temperature and the threshold is 20 °C, the data is divided into two groups: one where the temperature is above 20 °C and another where it's below
- Repeat the Process: This splitting process is repeated for each of the new groups (subsets) until a stopping criterion is met. A stopping criterion might be when all data points in a subset belong to the same category (e.g., all are suitable for bioenergy production), or when the subset is too small to split further
- Result: The result is a tree-like structure where each split represents a decision based on a feature and threshold. This structure helps predict outcomes, such as whether a certain area is suitable for bioenergy crops based on environmental conditions.

• Decision node: A decision node is a node in the decision tree that further splits into additional sub-nodes. It is an internal node that represents a decision based on a feature and a threshold value. Decision nodes are used to recursively partition the data into smaller subsets until the desired level of granularity is achieved as demonstrated below:

- A Decision Node: A decision node is an internal part of a decision tree that splits data into smaller groups. It's not the starting point (that's the root node) or the end point (those are leaf nodes), but rather a middle step
- Function of a Decision Node: Each decision node represents a decision based on a specific feature and a threshold value. For example, if you're analyzing data to predict bioenergy crop yields, a decision node might ask, "Is the soil pH above 6.5?" If yes, the data goes one way; if no, it goes another
- Recursive Partitioning: Decision nodes are used repeatedly to divide the data into smaller subsets until reaching the desired level of detail. This process is called recursive partitioning. It continues until each subset is small enough or until all data points in a subset belong to the same category (e.g., all are suitable for bioenergy production)
- Example in Bioenergy: Suppose using a decision tree to determine the best conditions for growing bioenergy crops. After the root node splits the data based on rainfall, a decision node might further split the subset with high rainfall based on another feature, such as sunlight hours. This process continues until you have subsets that are very specific, like "high rainfall and high sunlight," which might be ideal for certain bioenergy crops.

- Leaf node: A leaf node is a node in the decision tree that does not further split into additional sub-nodes. It represents a possible outcome or class for a given data point. Leaf nodes are the terminal nodes of the tree and are used to make predictions or classify data
- Branch: A branch is a subsection of the decision tree consisting of multiple nodes. It represents the path that the decision tree takes from the root node to a leaf node. Branches are used to visualize the decision-making process and to identify the features and conditions that lead to specific outcomes
- Pruning: Pruning is the process of removing sub-nodes of a decision tree to reduce its size and improve its performance. There are two main types of pruning: pre-pruning and post-pruning. Pre-pruning stops the tree from growing once it meets certain criteria, such as a maximum depth or a minimum number of samples per leaf. Post-pruning removes branches from a fully grown tree that do not provide significant power in classifying instances.

Decision Tree Construction Process Decision Tree Construction involves selecting the root node. This is often the most informative feature in the dataset. This is typically done by calculating the information gain or Gini impurity for each feature and selecting the one with the highest value. The selected feature is then used to split the data into two subsets based on a specific condition. For example, if the feature is a categorical variable, the data is split based on the categories. If the feature is continuous, the data is split based on a threshold value. The process of splitting continues recursively for each subset until a stopping criterion is met, such as a minimum number of samples per leaf or a maximum depth. Consider the simple example shown in Fig. 2.2. This is a decision tree to help decide if a certain coal can be combusted at a power plant. The requirement is that the %Carbon in the coal

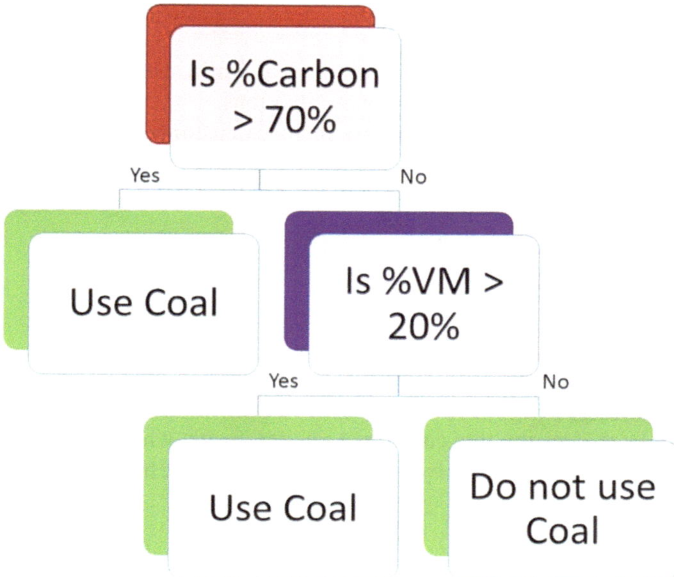

Fig. 2.2 Example of decision tree for classification

should be greater than 70% or the % Volatile Matter should be above 20%. The output of the decision tree is a decision on whether or not the coal can be used.

Decision trees turn complex rules into easy-to-follow steps. In the previous coal example, the tree simplifies the decision-making process for power plants by breaking it down into clear yes/no questions. This ensures consistent, data-driven choices about which coal to use, balancing efficiency and safety.

Limitations of Decision Trees

1. Overfitting:

   - Decision trees can become too complex and fit the training data too closely. This means they might not perform well on new, unseen data.
   - Example: Imagine a tree that perfectly predicts whether a customer buys a product based on many specific conditions in the training data. However, these conditions might not apply to new customers, leading to poor predictions.

2. Sensitivity to Small Changes:

   - Decision trees can be sensitive to small changes in the data. If the data changes slightly, the tree might look very different.
   - Example: If you add a few more data points or change some values slightly, the decision tree might split the data differently, leading to different predictions.

3. Limited Generalization:
   - If the training data doesn't represent the real-world scenario well, the decision tree might not generalize well to new situations.
   - Example: If your training data only includes customers from one region, the tree might not work well for customers from other regions.

Applications of Decision Trees

1. Classification Tasks:
   - Decision trees are great for classification tasks, like predicting whether a customer will buy a product based on their age, income, and past purchases.
   - Example: A decision tree can help a retailer decide which customers are most likely to buy a new product by analyzing their purchase history and demographic data.
2. Regression Tasks:
   - They are also used for regression tasks, such as forecasting stock prices based on historical data and economic indicators.
   - Example: A decision tree can predict stock prices by analyzing past trends, economic indicators, and company performance metrics.
3. Ensemble Methods:
   - Decision trees are a key part of ensemble methods like Random Forest. Random Forest combines many decision trees to improve accuracy and robustness.
   - How it Works: By creating multiple trees and combining their predictions, Random Forest reduces overfitting and improves the model's ability to generalize to new data. This makes it more reliable and accurate than a single decision tree.

2.1.2 Hyperparameters

Hyperparameters are configuration settings that are external to the machine learning model and cannot be directly extracted from the training data. Instead, they are set prior to the training process and govern the behavior and performance of the algorithm. Hyperparameters are distinct from model parameters, which are internal to the model and are adapted during training (e.g., weights in a neural network or coefficients in a linear regression). Hyperparameters play a crucial role in balancing the trade-off between bias and variance in a model, as they allow for the adjustment of model complexity to minimize error. By tuning these parameters, one can reduce bias by increasing model complexity, which helps the model fit the training data more closely, but this may also increase variance, causing the model to overfit (high variance) and perform poorly on unseen data. Conversely, reducing model complexity can decrease variance but may increase bias, leading to underfitting (high bias) and poor performance on both training and test data. Therefore, finding the optimal balance between bias and variance through hyperparameter tuning is essential for achieving high predictive accuracy and robust model performance.

Key Characteristics of Hyperparameters:

- Hyperparameters are chosen by the user before training begins. They are not updated during the training process
- They influence how the model learns, its capacity, and its ability to generalize to new data
- The choice of hyperparameters can affect significantly the model's accuracy, training speed, and convergence.

Examples of Hyperparameters:

1. Learning Rate: Controls the step size during optimization (e.g., gradient descent).
2. Number of Layers or Neurons in a Neural Network: Determines the architecture of the model.
3. Regularization Parameters (e.g., L1 or L2): Control overfitting by penalizing large weights.
4. Batch Size: The number of samples processed before the model is updated.
5. Number of Trees in a Random Forest: Defines the ensemble size.
6. Kernel Type in Support Vector Machines (SVM): Determines the shape of the decision boundary.

Selecting optimal hyperparameters is a critical step in building effective machine learning models. This process, known as hyperparameter tuning, involves systematically experimenting with different combinations of hyperparameters to find the best-performing configuration for a machine learning model. It requires evaluating how various settings of hyperparameters such as learning rate, regularization strength, batch size, and model complexity affect the model's performance on a validation dataset. The goal is to identify the optimal set of hyperparameters that maximizes the model's predictive accuracy, generalizability, and computational efficiency. Hyperparameter tuning can be performed manually through trial and error, or more efficiently using automated methods such as grid search, random search, or Bayesian optimization. Common techniques for hyperparameter tuning include:

- Grid Search: Exhaustively searching through a predefined set of hyperparameter values.
- Random Search: Randomly samples hyperparameter combinations, which can be more efficient than grid search for large search spaces.
- Bayesian Optimization: Uses probabilistic models to iteratively search for the best hyperparameters, often leading to faster convergence.
- Automated Tools: Leveraging libraries like Optuna, Hyperopt, or scikit-learn's GridSearchCV and RandomizedSearchCV.

2.2 Random Forest (RF)

The random forest algorithm consists of three main steps in ensemble learning (Prasad et al. 2006). The Random Forest algorithm is a powerful ensemble learning method that combines multiple decision trees to improve the accuracy and robustness of predictions. It starts with a process called bootstrap aggregation, or bagging, where the original dataset is randomly sampled with replacement to create multiple smaller datasets. This technique helps reduce bias and variance in the data, making the model more reliable, see Fig. 2.3.

For each of these smaller datasets, a decision tree is built. In a decision tree, each data point is evaluated at a decision node based on specific features. The data point is then directed to different branches of the tree depending on its characteristics. At the end of each branch, there's a leaf node that provides a prediction value.

In Random Forest, many decision trees are grown from their respective datasets. Each tree makes a prediction for a given input. The final prediction of the Random Forest model is calculated by averaging the predictions from all the individual decision trees. This averaging process helps to smooth out any irregularities and improves the overall performance of the model. Here's a step-by-step explanation of how it works:

1. Bootstrapping: Random Forest creates multiple subsets of the training data using a technique called bootstrapping. Each subset is created by randomly sampling observations from the original dataset with replacement.
2. Decision Tree Construction: For each bootstrap sample, a decision tree is grown. At each node of the tree, a subset of features is randomly selected, and the best split

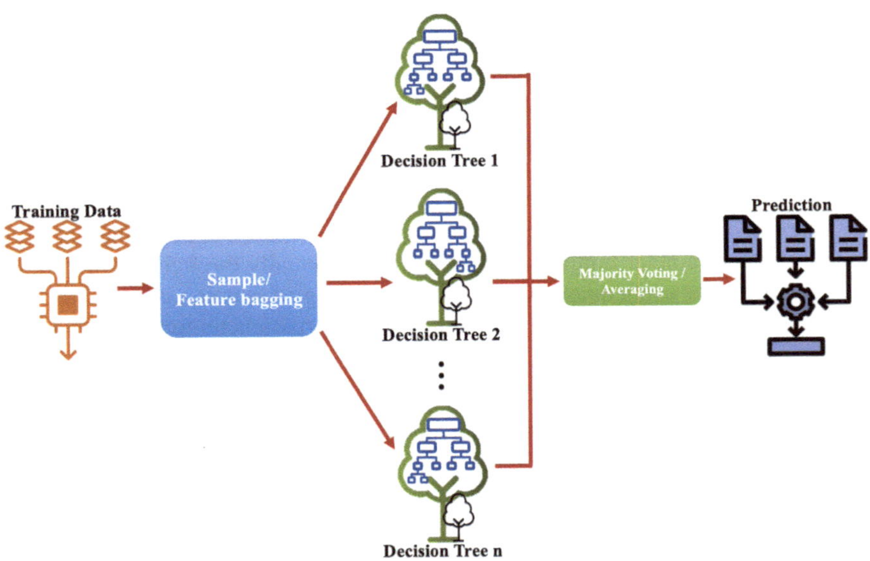

Fig. 2.3 Random forest algorithm model

is determined based on those features. This process is repeated until a stopping criterion is met, such as a maximum depth or a minimum number of samples per leaf.
3. Prediction: To make a prediction for a new data point, each decision tree in the forest makes a prediction. For classification tasks, the final prediction is the majority vote among all the trees. For regression tasks, the final prediction is the average of all the tree predictions.

The key aspects that make Random Forest effective are:

1. Bagging: By training each tree on a different subset of the data, Random Forest reduces the variance of individual trees and makes the overall model more robust to overfitting.
2. Feature Randomness: At each node, Random Forest considers only a random subset of features for splitting. This introduces additional randomness and diversity among the trees, further reducing the risk of overfitting.
3. Ensemble Strength: The combination of multiple decision trees in Random Forest leads to a more stable and accurate prediction compared to a single decision tree.

Advantages of Random Forest

1. Handles High-Dimensional Data: Random Forest can effectively handle datasets with a large number of features without feature selection.
2. Robust to Outliers and Noise: The ensemble nature of Random Forest makes it less sensitive to outliers and noise in the data.
3. Handles Missing Values: Random Forest can handle missing values in the data by using median/mode imputation or by creating additional branches in the trees.
4. Provides Feature Importance: Random Forest can estimate the importance of each feature in the prediction, which can be useful for feature selection and understanding the underlying relationships in the data.
5. Parallelizable: The training of individual trees in Random Forest can be parallelized, making it computationally efficient for large datasets.

RF is an advanced algorithm developed from multiple decision trees (DTs). Each DT receives a random subset of input data replacement (i.e., a bootstrap sample) and grows independently. Even if each DT in the RF is considered as a weak learner, when all the DTs make a prediction together, more proficiency and accuracy can be achieved. With a random subset data $d(\boldsymbol{x}, y)$, a DT recursively partitions the feature space, $\boldsymbol{x}$, as the samples with similar target values are grouped together. At node m, the data is represented by d_m with N_m samples. At each node in the decision tree, the data is split into two subsets based on a specific feature and threshold. For example, if we're considering a feature p and a threshold t_m at node m, the data can be divided into two subsets:

$$\begin{cases} d_m^{left} = \{(\boldsymbol{x}, y) | x_p \le t_m \\ d_m^{right} = d_m / d_m^{left} \end{cases} \tag{2.1}$$

Here, d_m^{left} includes all data points where the value of feature p is less than or equal to the threshold t_m, and d_m^{right} includes the remaining data points. To determine the best way to split the data, a loss function $H(\cdot)$, is used. This function measures how well the split improves the prediction. The loss function for a node m with N_m samples is given by

$$H(d_m) = \frac{1}{N_m} \sum_{y \in d_m} (y - \bar{y}_m)^2 \tag{2.2}$$

where $\bar{y}_m = \frac{1}{N_m} \sum_{y \in d_m} y$ is the average target value of all samples at node m. The algorithm selects the best split by minimizing a function that evaluates the quality of the split. This function, denoted as $G(\cdot)$, considers the loss for both subsets created by the split:

$$G(d_m, \theta) = \frac{N_m^{left}}{N_m} H(d_m^{left}(\theta)) + \frac{N_m^{right}}{N_m} H(d_m^{right}(\theta)) \tag{2.3}$$

Here, $\theta(p, t_m)$ represents the parameters of the split (feature p and threshold t_m), and N_m^{left} and N_m^{right} the number of samples in the left and right subsets, respectively. This is recursed for subsets d_m^{right} and d_m^{left} until the maximum depth (i.e., max_depth) allowed is reached. Once all decision trees are built, the final prediction from the Random Forest regression is calculated by averaging the outputs of all the individual decision trees:

$$f(\boldsymbol{x}) = \frac{1}{K} \sum_{k=1}^{K} DT_k(\boldsymbol{x}) \tag{2.4}$$

where K is the number of DTs ('n_estimators') in the RF. $f(x)$ is the final prediction for input $\boldsymbol{x}$. $DT_k(\boldsymbol{x})$ is the prediction from the $k-th$ decision tree. This results in considerable improvement in the generalization of the developed model (Phromphithak et al. 2021). Other main hyperparameters that are employed in both DT and RF algorithms are the minimum number of samples in splitting at each internal node ('min_samples_split') and the maximum leaf nodes in the tree ('max_leaf_nodes').

Applications of Random Forest Random Forest has been successfully applied in various domains, including:

1. Bioinformatics: Predicting protein-protein interactions, gene expression, and disease classification.
2. Image Recognition: Classifying images and detecting objects in images.
3. Fraud Detection: Identifying fraudulent transactions in financial data.
4. Recommendation Systems: Predicting user preferences and making personalized recommendations.
5. Ecology: Modeling species distributions and predicting habitat suitability.

Limitations of Random Forest

1. Interpretability: While Random Forest provides feature importance, it can be challenging to interpret the complex relationships between features and the target variable.

2. Memory Intensive: Training a large number of decision trees in Random Forest can be memory-intensive, especially for datasets with a large number of features.
3. Sensitive to Class Imbalance: Random Forest, like other tree-based algorithms, can be sensitive to class imbalance in the data, where one class is much more prevalent than others.
4. Overfitting: Although Random Forest is less prone to overfitting compared to individual decision trees, it can still overfit if the trees are grown too deep or if the number of trees is too large.

In conclusion, Random Forest is a powerful and versatile machine learning algorithm that combines multiple decision trees to make accurate and robust predictions. Its ability to handle high-dimensional data, missing values, and outliers, along with its feature importance capabilities, makes it a popular choice for various applications. However, it's essential to consider its limitations, such as interpretability and memory requirements, when applying it to specific problems.

2.2.1 Applications in Bioenergy

Random Forest has been successfully applied in various bioenergy-related tasks, such as estimating biomass potential, predicting energy yields, and optimizing energy production. For instance, in the context of agricultural biomass, Random Forest can be used to model the relationship between environmental and climatic factors and biomass production. This allows for more accurate predictions of biomass potential and more efficient resource allocation in the bioenergy sector.

2.3 Support Vector Machines

Support Vector Machines (SVMs) are a powerful machine learning algorithm used for classification, regression, and even outlier detection tasks. Support Vector Machines (SVMs) are a versatile and powerful supervised learning algorithm used for classification, regression, and outlier detection tasks. They are particularly effective for solving complex problems where data points are difficult to separate using simple boundaries. SVMs operate by finding the optimal decision boundary, called a hyperplane, that separates data points of different classes with the maximum margin—the distance between the hyperplane and the closest data points from each class. These closest points, known as support vectors, play a critical role in determining the position and orientation of the hyperplane.

Key Concepts of SVMs

1. Hyperplane:
   - A hyperplane is the decision boundary that separates different classes in the feature space. Its dimension depends on the number of features; for example, with two features, it is a line, and with three features, it becomes a plane.
   - SVMs aim to find the hyperplane that maximizes the margin between classes, ensuring better generalization to unseen data.
2. Support Vectors:
   - Support vectors are the data points closest to the hyperplane. They influence its position and orientation.
   - Removing or altering these points can change the hyperplane entirely, making them essential for building an SVM.
3. Kernel Trick:
   - When data is not linearly separable, SVMs use kernel functions to transform it into a higher-dimensional space where linear separation becomes possible.
   - Popular kernel functions include polynomial kernels, radial basis function (RBF) kernels, and sigmoid kernels.
4. Soft Margin:
   - For real-world datasets where perfect separation is not possible due to noise or overlapping classes, SVMs allow some misclassifications by maximizing a soft margin. This balances accuracy with robustness.

2.3.1 How SVMs Work

Support vector machine (SVM) is a supervised learning algorithm, originally developed for solving grouping problems (classification) and then generalized for regression tasks. This algorithm depends on the construction of a hyperplane or a set of hyperplanes in a high (or infinite) dimensional space for distinguishing groups of data, see Fig. 2.4. SVMs aim to find the optimal hyperplane that separates different classes of data with the maximum margin. The hyperplane is a fundamental concept that separates different classes of data. It is selected to maximize the distance from the closest data points of each class, known as the support vectors. The goal of an SVM is to find the maximum-margin hyperplane that separates classes with the largest possible margin.

Ideally, a good separation is when the hyperplane is positioned farthest from the closest training data points of any class (Garcia-Nieto et al. 2019). Given a feature dataset $\boldsymbol{x} = x_i \in R^p;\ i = 1, \ldots, n$ (with p features and n samples) and a target vector $y \in R^n$, SVR constructs a linear regression function in a transformed feature space (often using kernel methods):

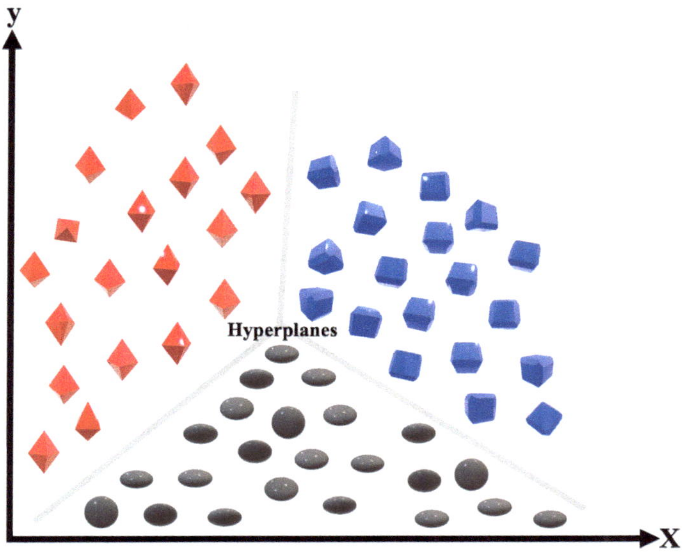

Fig. 2.4 Support vector machine algorithm model

$$f(\boldsymbol{x}) = \boldsymbol{w}^T \phi(\boldsymbol{x}) + b \tag{2.5}$$

where $\boldsymbol{w}$ and b are, respectively, weight vector determining the orientation of the regression hyperplane and bias terms shifting the hyperplane. $\Phi(x)$: A kernel-induced transformation that maps input features to a higher-dimensional space (e.g., polynomial or radial basis function kernels). The goal is to find the optimal w and b by solving the following constrained optimization problem:

$$\min_{w,b,\zeta,\zeta^*} \frac{1}{2}\boldsymbol{w}^T \boldsymbol{w} + C\sum_{i=1}^{n}\left(\zeta_i + \zeta_i^*\right) \tag{2.6}$$

Subject to constraints:

$$\{\begin{array}{c} y_i - \boldsymbol{w}^T\phi(x_i) - b \leq \varepsilon + \zeta_i \\ \boldsymbol{w}^T\phi(x_i) + b - y_i \leq \varepsilon + \zeta_i^* \\ \zeta_i, \zeta_i^* \geq 0 \end{array} \tag{2.7}$$

Explanation of Key Components

1. Objective Function

- Regularization Term $\frac{1}{2}\boldsymbol{w}^T\boldsymbol{w}$: Minimizes the complexity of the model (flatness of the regression hyperplane) to prevent overfitting. Smaller w values correspond to smoother functions.
- Penalty Term $C\sum_{i=1}^{n}\left(\zeta_i + \zeta_i^*\right)$: Penalizes deviations outside the ε-tube.

- C: Hyperparameter controlling the trade-off between model complexity and error tolerance.
- ζ_i, ζ_i^*: Slack variables representing how much a data point violates the ε-tube (errors above or below the tube, respectively).

2. Constraints ε-Tube: The first two constraints enforce that predictions $f(x_i)$ lie within an ε-margin of the true target y_i:

   - $y_i - f(x_i) \leq \varepsilon + \zeta_i$ (upper bound violation).
   - $f(x_i) - y_i \leq \varepsilon + \zeta_i^*$ (lower bound violation).
   - If a point lies inside the ε-tube (error $\leq \varepsilon$), no penalty is applied ($\zeta_i = \zeta_i^* = 0$).
   - Non-negativity: $\zeta_i, \zeta_i^* \geq 0$ ensures slack variables represent positive deviations only.

3. Hyperparameters ε: Defines the width of the tolerance tube.

   - A larger ε allows more deviation from the predicted function (simpler model).
   - A smaller ε forces the model to fit the data more tightly (risk of overfitting).

   C: Controls the penalty for errors outside the ε-tube.

   - A small $C < 1$ reduces penalties for slack variables, making the model more tolerant to noise (ideal for noisy datasets).
   - A large C strictly penalizes errors, potentially overfitting to training data.

The prediction function in SVR is derived using the Lagrangian dual technique, which transforms the original optimization problem into a form that leverages kernel functions for non-linear regression, (Garcia-Nieto et al. 2019). Here's a breakdown:

1. Prediction Formula The predicted output $f(\mathbf{x})$ for a new input $\mathbf{x}$ is calculated as

$$f(\mathbf{x}) = \sum_{[i=1]}^{n} (\alpha_i - \alpha_i^*) K(x_i, \mathbf{x}) + b \tag{2.8}$$

   - α_i, α_i^*: Lagrangian multipliers from the dual optimization problem. These determine the influence of each training sample x_i on the prediction.

     - α_i corresponds to deviations above the ε-tube (target > prediction).
     - α_i^* corresponds to deviations below the ε-tube (target < prediction).
     - Most α_i, α_i^* are zero; only support vectors (points outside the ε-tube) have non-zero values.

   - $K(x_i, \mathbf{x})$: The kernel function, which measures similarity between the new input $\mathbf{x}$ and the training sample x_i. It replaces explicit computations in high-dimensional space.
   - b: The bias term, determined during training.

- Role of the Kernel Function The kernel $K(x_i, x_j) = \phi(x_i)^T \phi(x_j)$ implicitly maps data to a higher-dimensional space where linear regression is possible. Common kernels include.

3. Why Use Kernels?

- Non-linear Regression: Kernels allow SVR to model complex relationships without explicitly transforming data into high dimensions (saves computation).
- Flexibility: Different kernels adapt to data structures (e.g., RBF for smooth curves, polynomial for periodic trends).

4. Hyperparameters in SVR

- γ (RBF Kernel):
  - Controls how far the influence of a single training example reaches.
  - Example: If $\gamma = 0.1$, distant points still affect the prediction; if $\gamma = 10$, only nearby points matter.
- p (Polynomial Kernel):
  - Controls the degree of the polynomial. Higher p captures more complex patterns but risks overfitting.
- C and ε:
  - C: Penalty for errors outside the ε-tube. Larger C = stricter fit (risk of overfitting).
  - ε: Width of the tolerance tube. Larger ε = more error allowed (simpler model).

2.3.2 Applications in Bioenergy

SVMs can be useful for bioenergy projects in several ways:

1. Biomass feedstock classification SVMs can be used to classify different types of biomass feedstocks based on their physical and chemical properties. This can help optimize the conversion process and improve the efficiency of bioenergy production.
2. Prediction of biofuel yields SVMs can be used to predict the yields of biofuels (e.g., bioethanol, biodiesel) based on various input parameters such as feedstock composition, pretreatment conditions, and fermentation conditions. This can help in process optimization and decision-making.

3. Fault detection and diagnosis SVMs can be used for fault detection and diagnosis in bioenergy plants, helping to identify and mitigate issues related to equipment failure, process upsets, or feedstock variability.
4. Optimization of process parameters SVMs can be used to optimize various process parameters in bioenergy production, such as temperature, pressure, and residence time, to maximize the efficiency and yield of the process.
5. Prediction of environmental impacts SVMs can be used to predict the environmental impacts of bioenergy production, such as greenhouse gas emissions, water usage, and land use change. This can help in the assessment of the sustainability of bioenergy projects.

Advantages of SVMs SVMs have several advantages that make them suitable for bioenergy applications:

1. Ability to handle high-dimensional data: SVMs can effectively handle data with a large number of features, which is common in bioenergy applications.
2. Robustness to overfitting: SVMs are less prone to overfitting compared to other machine learning algorithms, making them suitable for small datasets.
3. Flexibility in handling different types of data: SVMs can handle various types of data, including numerical, categorical, and text data.
4. Ability to handle non-linear relationships: SVMs can handle non-linear relationships by using kernel functions, which is important in many bioenergy applications.
5. Interpretability: SVMs provide insights into the importance of different features in the classification or regression task, which can be useful for understanding the underlying processes in bioenergy systems.

Support Vector Machines are a versatile and powerful machine learning algorithm that can be effectively applied to various bioenergy applications, including biomass feedstock classification, prediction of biofuel yields, fault detection and diagnosis, optimization of process parameters, and prediction of environmental impacts. The ability of SVMs to handle high-dimensional data, their robustness to overfitting, and their flexibility in handling different types of data make them a suitable choice for many bioenergy projects.

2.4 Kernel Ridge Regression

2.4.1 Regression

Regression is a versatile statistical method that extends beyond finance and investing to disciplines like bioenergy, where it plays a crucial role in understanding complex relationships between variables. In the context of bioenergy systems, regression analysis is used to determine the strength and character of the relationship between one dependent variable—such as biomass yield or biofuel production efficiency—and a series of independent variables, including environmental factors like temperature, precipitation, and soil quality. Also known as simple regression or ordinary least squares (OLS), linear regression is the most common form of this technique. It is used to identify associations between variables in bioenergy data, showing both the magnitude of these associations and their statistical significance, thereby distinguishing between meaningful relationships and those that might occur by chance.

Linear regression establishes a linear relationship between two variables based on a line of best fit, which can be particularly useful in bioenergy for modeling how changes in one variable impact another. For example, it can depict how a change in temperature affects biomass growth or how variations in sunlight exposure influence photosynthesis rates in bioenergy crops. The slope of the linear regression line defines how the change in one variable impacts a change in the other, while the y-intercept represents the value of one variable when the value of the other is zero. This could, for instance, show the baseline biomass yield when temperature is at a theoretical zero, providing insights into the intrinsic productivity of bioenergy crops.

In bioenergy systems, regression analysis is a powerful tool for statistical inference, enabling researchers to predict future outcomes based on past observations. By analyzing historical data on factors like climate conditions, soil health, and crop management practices, regression models can help predict biomass yields, optimize biofuel production processes, and inform strategic decisions about land use and resource allocation. This predictive capability is essential for developing sustainable and efficient bioenergy systems that can contribute effectively to renewable energy portfolios. Mathematically it can be represented as follows:

Simple linear regression:

$$Y = a + bX + u \tag{2.9}$$

Multiple linear regression:

$$Y = a + b_1 X_1 + b_2 X_2 + b_3 X_3 + \dots + b_t X_t + u \tag{2.10}$$

where: $Y =$ The dependent variable under enquiry
$X =$ The explanatory (independent) variable(s) used to predict or associate with Y
$a =$ The y-intercept
$b =$The slope of the explanatory variable(s)
$u =$ The regression residual or error term.

Take a regression model output in the form of $Y = 2.3 + (1.2)X_1 - 2.5(X_2) + 0.32$. This is a multiple linear regression that relates some variable Y with two explanatory variables X_1 and X_2. To interpret the model, the value of Y changes by 1.2 for every one-unit change in X_1, holding all else constant. That means controlling for X_2, X_1 has this observed relationship. Likewise, holding X_1 constant, every one-unit increase in X_2 is associated with a 2.5 decrease in Y. We can also note the y-intercept of 2.3, meaning that $Y = 2.3$ when X_1 and X_2 are both zero. The error term (residual) is 0.32.

Non-linear regression models also exist, but are far more complex. Regression analysis is a powerful tool for uncovering the associations between variables observed in data, but cannot easily indicate causation. It is used in several contexts in business, finance, and economics. For instance, it is used to help investment managers value assets and understand the relationships between factors such as commodity prices and the stocks of businesses dealing in those commodities.

2.4.2 Kernel Ridge Regression

Kernel Ridge Regression (KRR) is a powerful regression technique that combines ridge regression with the kernel trick, enabling the modeling of complex, non-linear relationships between features and target variables, see Fig. 2.5. It is particularly useful when linear regression models fail to capture the underlying patterns in data due to non-linear dependencies. Traditional Linear Regression assumes a linear relationship between features and the target variable, which can lead to poor performance when the true relationship is non-linear. KRR, by employing the kernel trick, maps data into a higher-dimensional space where non-linear relationships can be captured effectively. This allows KRR to model intricate patterns that linear regression cannot handle.

Key Concepts

1. Ridge Regression
   Ridge regression is a linear regression model that incorporates L2-norm regularization to prevent overfitting and stabilize the solution. Both KRR and ridge regression incorporate L2 regularization, which penalizes large coefficients and prevents overfitting. However, KRR extends this to non-linear models, ensuring that even complex functions remain generalizable and robust. The optimization problem for ridge regression is

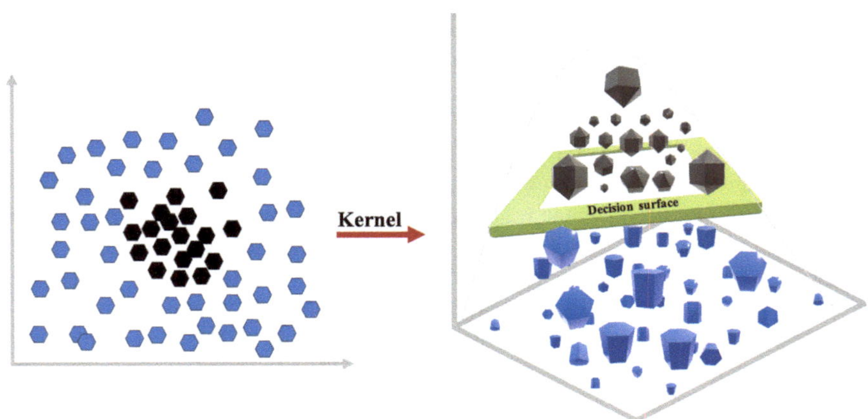

Fig. 2.5 Kernel ridge regression algorithm model

$$\min_{\mathbf{w},b} \frac{1}{2} \parallel \mathbf{w} \parallel^2 + C \sum_{[i=1]}^{N} (y_i - (\mathbf{w}^T \mathbf{x}_i + b))^2 \tag{2.11}$$

Here:

- **w**: Weight vector determining the orientation of the regression hyperplane.
- b: Bias term shifting the hyperplane.
- C: Regularization parameter controlling the trade-off between model complexity and fit to the data.

2. Kernel Trick
   The kernel trick allows KRR to handle non-linear relationships by implicitly mapping data into a higher-dimensional feature space without explicitly performing the transformation. This is achieved through kernel functions, which compute dot products in the higher-dimensional space based on original feature representations, see Table 2.1. The kernel trick avoids computationally expensive mappings while enabling effective modeling of non-linear patterns.

Kernel Ridge Regression Formulation KRR extends ridge regression to non-linear cases by learning a function in a Reproducing Kernel Hilbert Space (RKHS). The optimization problem can be written as

$$\hat{f} = \arg\min_{f \epsilon \mathcal{H}} \sum_{[i=1]}^{N} (y_i - f(x_i))^2 + \lambda \parallel f \parallel_{\mathcal{H}}^2 \tag{2.12}$$

Here:

- $\mathcal{H}$: RKHS induced by the chosen kernel.
- $\lambda > 0$: Regularization parameter balancing data fit and model complexity.

Using the representer theorem, the solution takes the form:

$$f(x) = \sum_{[i=1]}^{N} \alpha_i K(x_i, x) \tag{2.13}$$

Table 2.1 Set of three separate Kernel functions

Kernel	Formula	Behavior
Radial Basis Function (RBF)	$K(x_i, x_j) = \exp(-\gamma \parallel x_i - x_j \parallel^2)$	– γ controls the "spread" of influence – Small γ = broad influence – Large γ = localized influence
Polynomial	$K(x_i, x_j) = (1 + x_i^T x_j)^p$	– p is the polynomial degree – Higher p = more complex models
Linear	$K(x_i, x_j) = x_i^T x_j$	Simplest case; no transformation

where:

- α_i are coefficients learned during training.
- $K(x_i, x)$ is the kernel function evaluating similarity between training samples x_i and input x.

Optimization in KRR The functional optimization problem can be reformulated as a finite-dimensional quadratic program:

$$\hat{\alpha} = \arg\min_{\alpha \epsilon R^n} \parallel \mathbf{y} - \mathbf{K}\alpha \parallel_2^2 + \lambda \alpha^T \mathbf{K}\alpha \tag{2.14}$$

where:

- $\alpha = [\alpha_1, \ldots.., \alpha_N]^T$: Coefficient vector.
- $\mathbf{y} = [y_1, \ldots.y_N]^T$: Target vector.
- $\mathbf{K}$: Kernel matrix with entries $K(x_i, x_j)$.

The closed-form solution is

$$\hat{\alpha} = (\mathbf{K} + \lambda \mathbf{I})^{-1}\mathbf{y} \tag{2.15}$$

This enables efficient computation for medium-sized datasets.

Comparison with Support Vector Regression (SVR) While KRR and SVR share similarities in form, they differ in key aspects:

1. Loss Function:
   - KRR uses squared error loss.
   - SVR uses ε-insensitive loss, which ignores errors within a tolerance margin (ε).
2. Sparsity:
   - KRR learns a dense model using all training samples.
   - SVR learns a sparse model using only support vectors (samples outside the ε-tube).
3. Efficiency:
   - KRR fitting can be done in closed form and is faster for medium-sized datasets.
   - SVR scales better for large datasets due to sparsity.

Hyperparameters in KRR Key hyperparameters include:

1. Regularization Parameter (λ):
   - Controls trade-off between fitting the data and preventing overfitting.
   - Larger λ values result in simpler models with smoother predictions.

2. Kernel Parameters:

    - For RBF kernels: γ determines how far each training sample influences predictions.
    - For polynomial kernels: Degree (p) controls model complexity.

Applications Kernel Ridge Regression is widely used in scenarios requiring non-linear modeling, such as:

- Predicting complex physical phenomena (e.g., energy outputs in bioenergy systems).
- Modeling noisy relationships in datasets (e.g., sinusoidal trends with added noise).
- Handling structured data like moon-shaped or circular patterns.

Key Takeaways

- KRR combines ridge regression with kernel methods to handle non-linear relationships efficiently.
- It provides a dense solution that uses all training samples for predictions.
- Hyperparameters like regularization (λ) and kernel parameters (γ, p) must be carefully tuned for optimal performance.

By leveraging kernels and regularization, KRR offers an effective approach for tackling complex regression problems while maintaining computational efficiency for medium-sized datasets.

Disadvantages of KRR KRR involves mapping predictors into high-dimensional spaces, which can lead to computational complexity. This is particularly challenging for bioenergy projects that often require real-time or large-scale data analysis. Selecting appropriate tuning parameters (e.g., regularization factor and kernel parameters) is critical for KRR's performance. However, these parameters lack clear practical interpretation and often require time-consuming grid searches or cross-validation, which can result in suboptimal performance if not done carefully. The choice of kernel significantly affects the accuracy of predictions. If the kernel does not match the underlying data-generating process (e.g., using a polynomial kernel for non-polynomial data), forecast accuracy can suffer. This is problematic in bioenergy projects where data patterns might be complex and poorly understood.

KRR's performance depends on the signal-to-noise ratio in the data. Bioenergy data often includes noise from environmental factors, sensor inaccuracies, or incomplete datasets, which can reduce prediction reliability. While KRR can handle high-dimensional data, it may still face challenges when dealing with very large numbers of predictors or observations common in bioenergy systems. This can lead to over-fitting or reduced computational efficiency.

KRR models are less interpretable compared to simpler models like linear regression. Bioenergy projects often require transparent decision-making processes for stakeholder communication and regulatory compliance, which can be hindered by KRR's complexity. When the true data-generating process is unknown (as often occurs in bioenergy applications), KRR may estimate misspecified models, leading to reduced forecast quality.

These factors highlight why careful consideration is necessary before applying KRR to bioenergy projects, despite its flexibility and capability to model non-linear relationships effectively.

2.5 Gradient Tree Boosting

Gradient Tree Boosting (GTB), also known as Gradient Boosted Decision Trees (GBDT), is a powerful ensemble machine learning technique that combines multiple weak predictive models, typically decision trees, into a single, robust predictive model, (Friedman 2001; Chen and Guestrin 2016). It is widely used for regression, classification, ranking, and other supervised learning tasks due to its flexibility and ability to handle complex data relationships effectively. GTB builds its model iteratively by training weak learners (decision trees) sequentially. Each new tree is trained to correct the residual errors made by the ensemble of previous trees. This process ensures that the model progressively improves its predictions. The key steps include:

1. Residual Calculation: After each tree makes predictions, the residuals (errors) are computed as the difference between the actual target values and the predicted values
2. Gradient Descent Optimization: The algorithm uses gradient descent to minimize a predefined loss function (e.g., Mean Squared Error for regression or Cross-Entropy for classification). The gradients of the loss function guide the training of subsequent trees
3. Weak Learner Addition: A new tree is added to the ensemble, and its predictions are scaled by a learning rate to control their influence on the final model.

The loss function plays a central role in GTB by quantifying how well the model performs. Common loss functions include:

- Mean Squared Error (MSE) for regression tasks.
- Cross-Entropy for classification tasks. GTB optimizes this loss function iteratively in function space, using pseudo-residuals as targets for the weak learners. This functional gradient boosting approach allows GTB to generalize beyond traditional boosting methods.

Key Features

1. Shrinkage (Learning Rate): Shrinkage scales the contribution of each[] tree by a factor (learning rate, η), preventing overfitting and ensuring gradual improvement. Smaller learning rates require more trees but yield stronger models.
2. Tree Depth: The maximum depth of individual trees controls the level of interaction between variables. Shallow trees capture simple patterns, while deeper trees model more complex interactions.

3. Regularization: Techniques such as limiting tree depth, reducing learning rates, or applying early stopping help prevent overfitting while maintaining high predictive accuracy.

The prediction of Gradient Tree Boosting (GTB) models combines multiple weak learners, typically small decision trees, into an ensemble to improve accuracy. The ensemble's prediction for input features x_i is given by

$$f(x_i) = F_M(x_i) = \sum_{[m=1]}^{M} h(m)x_i \tag{2.16}$$

Here, h_m represents individual weak learners (decision tree regressors), and M denotes the total number of trees in the ensemble. GTB builds the model iteratively and greedily, updating predictions at each step as follows:

$$F_M(x_i) = F_{M-1}(x_i) + h_m(x_i) \tag{2.17}$$

Each new tree h_m is trained to minimize the sum of losses from the previous prediction $F_{M-1}(x_i)$. The optimization problem for h_m is

$$h(m) = \arg\min_h \sum_{[i=1]}^{N} l(y_i, F_{m-1}(x_i) + h(x_i)) \tag{2.18}$$

where $l(y_i, F_m)$ is a loss function that measures prediction error. For regression tasks using the least-squares method, GTB starts with an initial prediction F_0, which is the mean value of the target variable. Subsequent trees are trained on pseudo-residuals (errors between actual and predicted values), progressively refining predictions. For example:

- The first tree predicts residuals using the target mean.
- The second tree updates predictions to $F_1(x_i) = F_0(x_i) + h_1(x_i)$.
- The process repeats for M iterations, adding trees until further improvement becomes negligible.

A learning rate scales each tree's contribution to control overfitting and improve generalization. This iterative procedure ensures that GTB models achieve high accuracy while maintaining robustness against overfitting.

Disadvantages of Gradient Tree Boosting Gradient Tree Boosting (GTB) is a widely used machine learning algorithm, but it has several disadvantages that can impact its application in bioenergy projects. GTB can overfit noisy data, especially when the number of trees or iterations is too high. Training GTB models is computationally expensive because trees are built sequentially rather than in parallel. This makes GTB less efficient for large-scale bioenergy projects with extensive datasets or real-time prediction requirements. GTB requires careful tuning of hyperparameters like learning rate, number of trees, and tree depth. Improper tuning can lead to poor performance or excessive training time.

GTB models are less interpretable compared to simpler models like linear regression. The sequential training process prevents parallelization, making GTB slower compared to algorithms like Random Forests that allow parallel training. GTB relies heavily on high-quality data. Missing values, outliers, or inconsistent measurements can negatively impact model accuracy and reliability. Generating probabilistic predictions with GTB can be computationally intensive and complex. While GTB offers high accuracy and flexibility for modeling non-linear relationships, these challenges highlight the need for careful implementation and consideration of alternative methods.

2.6 eXtreme Gradient Boosting (XGB) Algorithm

XGBoost, short for eXtreme Gradient Boosting, is an advanced machine learning algorithm designed to optimize the traditional Gradient Boosting framework. It builds predictive models by combining multiple weak learners, typically decision trees, into a strong ensemble, (Alabdrabalnabi et al. 2022; Huang et al. 2022; Katongtung et al. 2022). The algorithm works iteratively, where each new tree is trained to correct the errors made by the previous ones, see Fig. 2.6. This process minimizes a loss function, such as Mean Squared Error for regression or Log Loss for classification, to improve overall performance.

What sets XGBoost apart is its efficiency and scalability. It incorporates features like parallel processing, regularization (L1 and L2 penalties), and a sparsity-aware system for handling missing data. These enhancements make it faster and more robust than standard Gradient Boosting implementations while reducing the risk of overfitting. XGBoost is widely used in various applications due to its ability to handle large datasets and deliver high accuracy in predictive modeling tasks.

XGBoost (eXtreme Gradient Boosting) is an advanced implementation of Gradient Tree Boosting with several enhancements that improve performance, scalability, and efficiency. Below are the key differences:

1. Regularization:
   XGBoost includes explicit regularization techniques, such as L1 and L2 penalties, to control overfitting. Regular GTB implementations often lack these features, making XGBoost more robust for complex datasets.
2. Handling Missing Values:
   XGBoost efficiently handles missing values by learning optimal splits for missing data during training. Regular GTB implementations may require preprocessing to deal with missing values.
3. Parallelization:
   XGBoost supports parallel processing by building tree nodes simultaneously within each depth level, significantly reducing training time. Regular GTB algorithms are sequential and cannot parallelize this process.

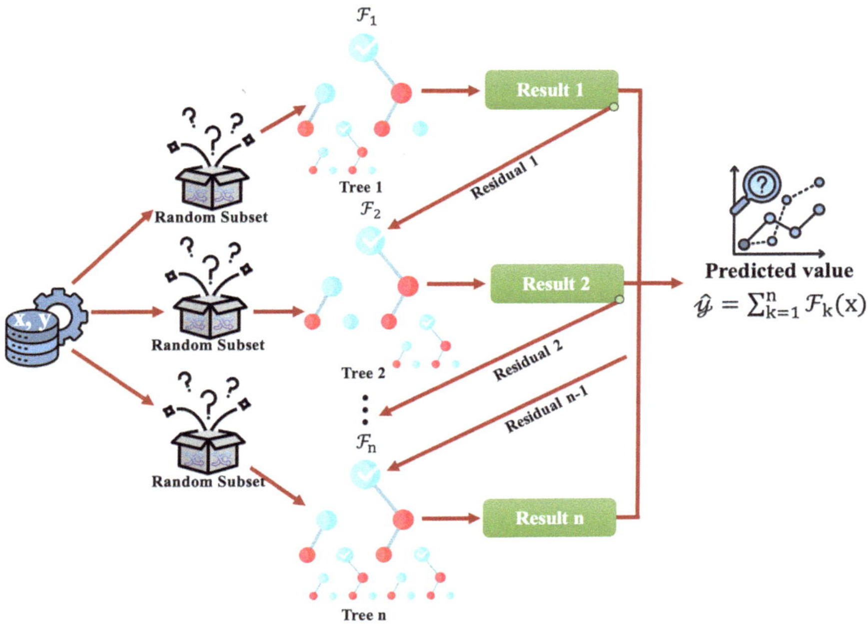

Fig. 2.6 Extreme gradient boosting algorithm model

4. Speed and Memory Optimization:
   XGBoost uses optimized data structures and cache-aware computations to improve speed and memory utilization. It also supports sparse matrices for handling large datasets efficiently. Regular GTB implementations are generally slower and less memory-efficient.
5. Row and Column Sampling:
   XGBoost introduces randomization through row sampling (subsets of training data) and column sampling (subsets of features), reducing overfitting and improving generalization. Regular GTB typically uses the entire dataset for training each tree.
6. DART (Dropouts Meet Additive Regression Trees):
   XGBoost implements DART, a dropout regularization technique that randomly drops trees during training to reduce overfitting further. This is not available in standard GTB.
7. Scalability:
   XGBoost scales well with large datasets due to its distributed learning capabilities across multicore machines or clusters. Regular GTB struggles with scalability for very large datasets.

Thanks to regularization, parallelization, and efficient handling of sparse data, XGBoost often achieves better predictive accuracy than regular GTB implementations. Parallel node construction and optimized computations make XGBoost significantly faster than regular GTB, especially for large datasets. Regularization techniques (L1/L2 penalties, DART) and sampling methods help XGBoost avoid overfitting more effectively than standard GTB. Features like sparse matrix support and distributed learning make XGBoost ideal for handling large-scale bioenergy datasets or other real-world applications requiring scalability. Automatic handling of missing values simplifies preprocessing steps compared to regular GTB, making it more practical for bioenergy projects where data gaps are common.

Ideally, XGB makes predictions of a target variable via a set of simple, weak models, which are generally decision trees for regression. In each training progress, a new decision tree is added, in which it is grown from residuals or errors of prior decision trees, and then, the new tree is assembled with previous ones to make a prediction. To decide adding a new tree, a gradient descent algorithm (sometimes called gradient boosting) is applied to minimize a regularized objective function, consisting of a (differentiable) convex loss function and a penalty term for model complexity (Chen and Guestrin 2016). In other words, XGB employs a gradient boosting technique to grow/develop a sequence of decision trees to make a prediction.

Mathematically, for a given $\{\mathbf{x}, \mathbf{y}\}$ with n datapoints and p features, a prediction, $f(x)$, is made from a tree ensemble model of M additive functions:

$$f(x) = \sum_{[m=1]}^{M} f_m(x), \qquad f_m \in F \tag{2.19}$$

where f_m is an independent decision tree in the space of regression trees, F. Compared to other GB algorithms, XGB has a unique ability, in which the loss function, $l(f(\boldsymbol{x}), y)$, is approximated by second-order Taylor expansion, while the objective function, $\mathcal{L}$, includes a regularization term, $\Omega(f_k)$:

$$\mathcal{L} = \sum_{i} l(f_i(x), y_i) + \sum_{k} \Omega(f_k) \tag{2.20}$$

The objective function ($\mathcal{L}$) of XGB at the t-th iteration is

$$\mathcal{L}^t = \sum_{[i=1]}^{p} l(f_t(x_i) + f_{t-1}(x_i), y_i) + \sum_{[m=1]}^{t} \Omega(f_m) \tag{2.21}$$

$$\Omega = \gamma T + \frac{1}{2}\lambda\|w\|^2 \tag{2.22}$$

where l is the loss function, Ω is the penalty term, γ and λ are hyperparameters of the penalty term, and T is the number of leaves in the decision tree with leaf weights, w. With Taylor's expansion, the $\mathcal{L}$ can be simplified to

$$\mathcal{L}^t = \sum_{[i=1]}^{p} \left[f_t(x_i)g_i + \frac{1}{2}(f_t(x_i))^2 h_i \right] + \gamma T + \frac{1}{2}\lambda \sum_{[j=1]}^{T} w_j^2 \tag{2.23}$$

where g_i and h_i are the first and second derivatives of the l, respectively. In practice, as a greedy algorithm, XGB starts with a single leaf and then raises branches into the tree. The feature candidates of splits are evaluated by

$$\mathcal{L}^{split} = \frac{1}{2}\left[\frac{\left(\sum_{i=I_R} g_i\right)^2}{\sum_{i=I_R} h_i + \lambda} + \frac{\left(\sum_{i=I_L} g_i\right)^2}{\sum_{i=I_L} h_i + \lambda} - \frac{\left(\sum_{i=I} g_i\right)^2}{\sum_{i=I} h_i + \lambda}\right] - \gamma \tag{2.24}$$

where I_R and I_L are the subsets of data at the right and left nodes after the split, respectively, and $I = I_R + I_L$. For hyperparameters, the higher γ and λ values are, the more conservative the XGB algorithm will be. Another important hyperparameter is 'learning_rate', which is applied after each boosting iteration to shrink the feature weights and prevent overfitting. Other hyperparameters of XGB can be found in the literature (Chen and Guestrin 2016).

Disadvantages of Applying XGBoost While XGBoost is a powerful algorithm for predictive modeling, it has several disadvantages that can impact its application in bioenergy projects. XGBoost models can be complex and difficult to interpret, especially when dealing with deep trees or numerous levels. This complexity makes it challenging to understand how different features interact, which is crucial for decision-making in bioenergy projects. Although XGBoost is efficient for many datasets, its level-wise tree growth strategy can be slower than other algorithms like LightGBM for very large datasets. This can lead to increased computational time and resource usage, which might be problematic for projects requiring rapid model updates or real-time predictions. The level-wise growth technique in XGBoost can cause memory issues when dealing with large feature sets or datasets, potentially limiting its application in memory-constrained environments. Despite built-in regularization techniques, XGBoost can still overfit if not properly tuned, especially with noisy or complex bioenergy datasets. This requires careful hyperparameter optimization to avoid capturing irrelevant patterns.

While XGBoost excels at modeling non-linear relationships, it might struggle with extremely complex or non-linear interactions present in some bioenergy systems. In such cases, other models like neural networks might be more effective. XGBoost has several hyperparameters that need tuning for optimal performance, such as learning rate, tree depth, and number of trees. This tuning process can be time-consuming and requires significant expertise, which might be a challenge for bioenergy projects with limited resources or tight deadlines. Like many machine learning algorithms, XGBoost's performance is heavily dependent on data quality. Bioenergy datasets often contain missing values, outliers, or inconsistent measurements, which can negatively impact model accuracy unless properly addressed.

These challenges highlight the need for careful consideration and optimization when applying XGBoost to bioenergy projects.

2.6.1 SHAP Analysis

SHAP (SHapley Additive exPlanations) is a powerful technique for interpreting machine learning models by explaining how input features contribute to the model's predictions. It is based on concepts from cooperative game theory, specifically Shapley values, which measure the contribution of each "player" (in this case, a feature) to the overall outcome of a "game" (the model's prediction), (Shapley et al. 1953).

In SHAP analysis, each feature is assigned a value that quantifies its impact on the prediction for a given data instance. These SHAP values decompose the model's output into contributions from each feature, allowing you to understand both local and global behavior:

- Local Analysis: Explains individual predictions by showing how each feature's value contributes to the specific prediction.
- Global Analysis: Aggregates SHAP values across all instances to reveal overall feature importance and interactions.

For any prediction $f(x)$, SHAP ensures that the model output can be expressed as

$$f(x) = base\ value + \sum SHAP\ values \tag{2.25}$$

The base value represents the average model output across all data points, while SHAP values indicate how much each feature pushes the prediction away from this base value. For example, in regression tasks, the base value could be the mean target value, and for classification tasks, it might represent the prevalence of a specific class.

SHAP has several desirable properties:

1. Additivity: The sum of SHAP values equals the difference between the model's prediction and the base value.
2. Consistency: If a feature's contribution increases in a new model, its SHAP value will not decrease.
3. Local Accuracy: SHAP values precisely explain individual predictions.

Practically, SHAP is implemented using Python libraries like https://pypi.org/project/shap/ by Lundberg et al. (2020), which provide tools for visualizing feature contributions through summary plots, dependence plots, and force plots. These visualizations make it easier to interpret complex models like XGBoost, offering insights into which features drive predictions and how they interact.

In bioenergy projects or other domains where interpretability is crucial, SHAP can help stakeholders understand how input variables (e.g., weather conditions or biomass characteristics) influence predictions, fostering trust in machine learning models and aiding in decision-making. SHAP can help interpret/explain how input features affect the predictions of the XGB model.

SHAP calculates significance values (called SHAP values) for each input feature in a prediction, ensuring three key properties:

1. Local Accuracy: The sum of feature contributions equals the model's prediction. Mathematically, the explanation model $G(\hat{A}')$ satisfies

$$f(x) = G(\hat{A}') = \phi_0 + \sum_{j=1}^{N} \phi_j \hat{A}'_j \tag{2.26}$$

   Here, ϕ_0 is the baseline prediction (e.g., the mean prediction when no features are used), and ϕ_j represents the contribution of feature j.
2. Missingness: Features that are absent ($A'_j = 0$) have no influence on the prediction, resulting in $\phi_j = 0$.
3. Consistency: If a feature's contribution increases or remains unchanged in a new model, its SHAP value should not decrease.

The formula for SHAP values is derived from cooperative game theory and ensures fairness in distributing the "payout" (the prediction) among features:

$$\phi_j(f, x) = \sum_{\hat{A}'_z \subseteq \hat{A}'} \frac{\mid \hat{A}'_z \mid !(N- \mid \hat{A}'_z \mid -1)!}{N!} [f(x)(\hat{A}'_z) - f(x)(\hat{A}'_{zj})] \tag{2.27}$$

Here:

- $\mid \hat{A}'_z \mid$ is the number of non-zero entries in $\hat{A}'_z$.
- $f(x)(\hat{A}'_z)$ represents the model's prediction using subset $\hat{A}'_z$.
- $f(x)(\hat{A}'_{zj})$ represents predictions excluding feature j.

SHAP values are particularly useful for both local interpretability (explaining individual predictions) and global interpretability (understanding overall feature importance across a dataset). This makes SHAP an effective tool for analyzing machine learning models, including complex ones like XGBoost, while maintaining transparency and trustworthiness.

Key Issues with SHAP Analysis SHAP often assumes that features are independent when calculating their contributions to predictions. In reality, features in datasets, especially those related to bioenergy (e.g., weather conditions, biomass properties), are frequently correlated. Ignoring these dependencies can lead to inaccurate or misleading attributions. SHAP explains feature importance in terms of the model's predictions but does not assess causal relationships. For example, a feature might appear important to the model but may not have a real-world causal impact on the target variable. This limitation is critical in bioenergy projects where understanding causality (e.g., the effect of temperature on energy output) is essential. Users may misinterpret SHAP values by assuming they represent absolute importance or causal effects, leading to incorrect conclusions. Additionally, biases can arise when users create narratives that align with their expectations rather than the actual data-driven insights.

Calculating SHAP values can be computationally expensive, especially for large datasets or complex models like XGBoost. KernelSHAP, in particular, is slow and impractical for large-scale analyses requiring explanations for many instances. SHAP assigns equal importance to all marginal contributions of features, regardless of whether larger or smaller feature sets are more informative. This uniform weighting can lead to suboptimal attributions in certain scenarios, as demonstrated by studies proposing alternatives like WeightedSHAP. SHAP values reflect how features contribute to the model's predictions but are only as reliable as the model itself. If the underlying machine learning model is poorly trained or biased, SHAP explanations will also be flawed.

Considerations for Bioenergy Projects In bioenergy projects, these limitations highlight the need for careful implementation and interpretation of SHAP analysis:

- Address feature dependencies through preprocessing or alternative methods.
- Avoid using SHAP for causal inference; complement it with domain expertise or causal analysis techniques.
- Ensure robust model training and validation to improve the reliability of SHAP explanations.

By understanding these limitations and applying best practices, users can leverage SHAP effectively while minimizing potential pitfalls.

2.7 Data Standardization

Standardization and normalization are critical preprocessing steps in machine learning to ensure that datasets are in a suitable format for algorithms to learn effectively. These processes help eliminate biases caused by varying scales or distributions of input features, leading to better model performance. In the case described, the Yeo-Johnson transformation was applied to standardize the dataset.

The Yeo-Johnson transformation, introduced by Yeo and Johnson (2000), is a non-linear method designed to improve the normality or symmetry of a dataset's distribution. It is particularly useful for datasets containing both positive and negative values, unlike the Box-Cox transformation, which only works with positive values. The transformation applies a power function to the data, where the optimal parameter (λ) is estimated to minimize skewness or improve normality. This makes it suitable for datasets with complex distributions often encountered in real-world applications.

After applying the Yeo-Johnson transformation, the data can be further standardized by centering (subtracting the mean) and scaling (dividing by the standard deviation). This ensures that the transformed dataset approximates a standard normal distribution, with a mean of 0 and a standard deviation of 1. The combination of transformation and standardization ensures that machine learning algorithms can process the data more effectively, as many algorithms (e.g., gradient-based methods) perform better when features have similar scales and distributions.

In practice, tools like R's bestNormalize package or Python libraries can automate this process. The transformation can also be inverted if needed, allowing predictions or outputs to be mapped back to their original scale. By using Yeo-Johnson's method, preprocessing becomes more robust for datasets with mixed-value ranges, making it ideal for complex applications such as bioenergy modeling or other domains requiring precise data preparation.

Yeo-Johnson's non-linear transformation (Yeo and Johnson 2000) is shown in Eq. 2.28:

$$\widetilde{x_l} = \left\{ \begin{array}{c} \left[(\widetilde{x_l}+1)^{\lambda}-1\right]/\lambda, \quad \text{If } \lambda \neq 0, \widetilde{x_l} \geqslant 0 \\ \ln(\widetilde{x_l}+1), \quad \text{If } \lambda = 0, \widetilde{x_l} \geqslant 0 \\ -\left[(-\widetilde{x_l}+1)^{2-\lambda}-1\right]/(2-\lambda), \quad \text{If } \lambda \neq 2, \widetilde{x_l} < 0 \\ -\ln(-\widetilde{x_l}+1), \quad \text{If } \lambda = 2, \widetilde{x_l} < 0 \end{array} \right\} \tag{2.28}$$

where $\widetilde{x_l}$ is the transformed feature data from the raw feature data x_i and λ is the transformation parameter, which is determined by maximum possibility estimation. The subscript i runs from 1 to N which is the number of data in each feature. Then, all input values were normalized as

$$x_i^* = \frac{x_i - \mu}{\sigma} \tag{2.29}$$

where x_i^* is the normalized value of each input variable x_i, while μ and σ are the mean value and standard deviation of each input variable respectively.

To assess the accuracy of model predictions, root means square error ($RMSE$) and regression coefficient (R^2) was used. $RMSE$ is a standard mathematical perspective outlook to measure errors of predicted values compared to their real values, while R^2 is widely known as a statistical measurement for any mathematical model performance, showing the strength of the relationship between the dependent and independent variables. Conceptually, the lower the $RMSE$ and the higher the R^2, the greater is the model accuracy (Tang et al. 2020). These $RMSE$ and R^2 parameters were calculated, as Eqs. (2.30) and (2.31), respectively:

$$R_t^2 = 1 - \frac{\Sigma_{i=1}^{n}\left(Y_{pred,i}^t - Y_{exp,i}^t\right)^2}{\Sigma_{i=1}^{n}\left(Y_{exp,i}^t - Y_{exp,avg}^t\right)^2} \tag{2.30}$$

$$RMSE_t = \sqrt{\frac{\Sigma_{i=1}^{n}\left(Y_{pred,i}^t - Y_{exp,i}^t\right)^2}{n}} \tag{2.31}$$

where $Y_{pred,i}^t$ and $Y_{exp,i}^t$ are the predicted and experimental (real) values of target t, respectively, while $Y_{exp,avg}^t$ is the average of all experimental values of target t. n is the total number of data points, and $RMSE_t$ represents the root mean squared error of target t.

2.7.1 Feature Correlation

A good dataset for machine learning is one where the input features are strongly correlated with the target variable (output) but exhibit minimal correlation with each other, (Senliol et al. 2008). This ensures that the features contribute uniquely to the model's predictions without introducing redundancy or multicollinearity, which can degrade model performance.

To measure the linear relationship between two variables, Pearson's correlation coefficient is often used. It is a simple, filter-based method that quantifies the strength and direction of a linear relationship between two continuous variables. The Pearson correlation coefficient ranges from –1 to 1:

- A value close to 1 indicates a strong positive correlation (as one variable increases, so does the other).
- A value close to –1 indicates a strong negative correlation (as one variable increases, the other decreases).
- A value near 0 suggests no linear relationship between the variables.

In practical terms, Pearson's correlation can be applied during feature selection in machine learning to:

1. Identify features that are strongly correlated with the target variable, as these are likely to be more predictive.
2. Detect and remove features that are highly correlated with each other (multicollinearity), as such redundancy can cause instability in model training and reduce interpretability.

For example, in a dataset with features like temperature, humidity, and energy output in a bioenergy project, Pearson's correlation can help determine which features have the strongest linear relationship with energy output (target). At the same time, it can identify if temperature and humidity are highly correlated with each other, prompting the removal of one to prevent redundancy.

However, Pearson's correlation only measures linear relationships. If relationships between variables are non-linear, alternative methods like Spearman's rank correlation or mutual information may be more appropriate. Pearson's correlation measures the linear dependence between any two variables, as follows, Eq. (2.32):

$$r = \frac{\Sigma_{i=1}^{N} x_i * \Sigma_{j=1}^{N} x_j}{\left[\left(\Sigma_{i=1}^{N} x_i^2 \right)^{1/2} * \left(\Sigma_{j=1}^{N} x_j^2 \right)^{1/2} \right]} \tag{2.32}$$

where r is the Pearson correlation coefficient of evaluated features x_i and x_j having properties of zero mean and unit variance. The correlation coefficient can be both positive or negative values indicating a correlation direction of those considered features. If any two features have a high value of the correlation coefficient, one of them should be excluded from the dataset. The correlation coefficient's significant level test (p-value) can be calculated by a two-tailed t distribution with $N - 2$ degrees of freedom using Eq. 2.33.

$$t = \frac{r\sqrt{n-2}}{\sqrt{1-r^2}} \tag{2.33}$$

where n is numbers of data points and r is the PCC between any two variables.

References

Alabdrabalnabi A, Gautam R, Sarathy SM (2022) Machine learning to predict biochar and bio-oil yields from co-pyrolysis of biomass and plastics. Fuel 328:125303. ISSN 0016-2361. https://doi.org/10.1016/j.fuel.2022.125303. https://www.sciencedirect.com/science/article/pii/S001623612202138X

Chen T, Guestrin C (2016) Xgboost: a scalable tree boosting system. In: Proceedings of the 22nd ACM SIGKDD international conference on knowledge discovery and data mining, KDD '16, pp 785–794. Association for Computing Machinery, New York, NY, USA. ISBN 9781450342322. https://doi.org/10.1145/2939672.2939785

Friedman JH (2001) Greedy function approximation: a gradient boosting machine. Ann Stat 29(5):1189–1232

Garcia-Nieto P, Garcia-Gonzalo E, Paredes-Sanchez J, Bernardo-Sanchez A, Menendez-Fernandez M (2019) Predictive modelling of the higher heating value in biomass torrefaction for the energy treatment process using machine learning techniques. Neural Comput Appl 31(12):8823–8836. https://doi.org/10.1007/s00521-018-3870-x

Huang N, Gao K, Yang W, Pang H, Yang G, Wu J, Zhang S, Chen C, Long L (2022) Assessing sediment organic pollution via machine learning models and resource performance. Bioresour Technol 361:127710. ISSN 0960-8524. https://doi.org/10.1016/j.biortech.2022.127710. https://www.sciencedirect.com/science/article/pii/S0960852422010392

Jentzer JC, Kashou AH, Murphree DH (2023) Clinical applications of artificial intelligence and machine learning in the modern cardiac intensive care unit. Intell Based Med 7:100089. ISSN 2666-5212. https://doi.org/10.1016/j.ibmed.2023.100089. https://www.sciencedirect.com/science/article/pii/S2666521223000030

Katongtung T, Onsree T, Tippayawong N (2022) Machine learning prediction of biocrude yields and higher heating values from hydrothermal liquefaction of wet biomass and wastes. Bioresour Technol 344:126278. ISSN 0960-8524. https://doi.org/10.1016/j.biortech.2021.126278. https://www.sciencedirect.com/science/article/pii/S0960852421016205

Lundberg SM, Lee S-I (2017) A unified approach to interpreting model predictions. In: Guyon I, Luxburg UV, Bengio S, Wallach H, Fergus R, Vishwanathan S, Garnett R (eds) Advances in neural information processing systems, vol 30. Curran Associates, Inc. https://proceedings.neurips.cc/paper_files/paper/2017/file/8a20a8621978632d76c43dfd28b67767-Paper.pdf

Lundberg SM, Erion G, Chen H, DeGrave A, Prutkin JM, Nair B, Katz R, Himmelfarb J, Bansal N, Lee S-I (2020) From local explanations to global understanding with explainable AI for trees. Nat Mach Intell 2:56–67

Onsree T, Tippayawong N (2021) Machine learning application to predict yields of solid products from biomass torrefaction. Renew Energy 167:425–432. ISSN 0960-1481. https://doi.org/10.1016/j.renene.2020.11.099. https://www.sciencedirect.com/science/article/pii/S0960148120318516

Pedregosa F, Varoquaux G, Gramfort A, Michel V, Thirion B, Grisel O, Blondel M, Prettenhofer P, Weiss R, Dubourg V, Vanderplas J, Passos A, Cournapeau D, Brucher M, Perrot M, Duchesnay E (2011) Scikit-learn: machine learning in Python. J Mach Learn 12:2825–2830

Phromphithak S, Onsree T, Tippayawong N (2021) Machine learning prediction of cellulose-rich materials from biomass pretreatment with ionic liquid solvents. Bioresour Technol 323:124642. ISSN 0960-8524. https://doi.org/10.1016/j.biortech.2020.124642. https://www.sciencedirect.com/science/article/pii/S0960852420319167

Prasad AM, Iverson LR, Liaw A (2006) Newer classification and regression tree techniques: bagging and random forests for ecological prediction. Ecosystems 9(2):181–199. ISSN 1435-0629. https://doi.org/10.1007/s10021-005-0054-1

Senliol B, Gulgezen G, Yu L, Cataltepe Z (2008) Fast correlation based filter (FCBF) with a different search strategy. In: 2008 23rd international symposium on computer and information sciences, pp 1–4. https://doi.org/10.1109/ISCIS.2008.4717949

Shapley LS et al (1953) A value for n-person games

Tang Q, Chen Y, Yang H, Liu M, Xiao H, Wu Z, Chen H, Naqvi SR (2020) Prediction of bio-oil yield and hydrogen contents based on machine learning method: effect of biomass compositions and pyrolysis conditions

Yeo I-K, Johnson RA (2000) A new family of power transformations to improve normality or symmetry. Biometrika 87(4):954–959. ISSN 0006-3444. https://doi.org/10.1093/biomet/87.4.954

Zhu X, Li Y, Wang X (2019) Machine learning prediction of biochar yield and carbon contents in biochar based on biomass characteristics and pyrolysis conditions. Bioresour Technol 288(121527). https://doi.org/10.1016/j.biortech.2019.121527

Chapter 3
Pretreatment with Ionic Liquids

Abstract Ionic liquid solvents (ILSs) have emerged as effective agents for processing biomass, resulting in cellulose-rich materials (CRMs) that can be utilized across various fields. However, accurately predicting the properties of CRMs and deciphering the intricate interactions within this system can be quite complex. In this chapter, we explore the application of machine learning algorithms to forecast CRM characteristics, with a particular emphasis on the cellulose enrichment factor and solid recovery. The Cellulose Enrichment Factor (CEF) is a measure of how effectively a pretreatment or processing method increases the relative concentration of cellulose in the material compared to its original state. Solid Recovery (SR) refers to the proportion of the solid material that remains after a pretreatment or processing step, relative to the initial amount of raw material. The machine learning algorithms (MLA) analyzed datasets comprising of 23 features, which encompass biomass properties, operational conditions, the specific ILS's, and catalyst details. Among the various algorithms evaluated, the random forest model proved to be the most reliable, achieving a root mean square error (RMSE)of 0.22 and an R^2 value of 0.94. This indicates a strong correlation between the predicted and actual values. Notably, biomass properties and ILS operating conditions were identified as the most influential factors, contributing 80% to the CEF and 60% to SR. To aid in visualizing and interpreting the relationships between these significant features and CRM properties, one-way and two-way partial dependence plots are presented. These findings potentially useful for designing ILSs or enhancing process efficiency.

Keywords Machine learning algorithms · Ionic liquid solvents · Cellulose-rich materials · Bioenergy

3.1 Introduction to Ionic Liquids

Lignocellulosic biomass is composed of three primary components: cellulose, hemicellulose, and lignin. Generally, biomass contains about 40–60% cellulose by weight, 15–30% hemicellulose, and 10–25% lignin. However, the exact proportions can vary based on the specific type of biomass. These components are intricately linked,

N. Tippayawong et al., *Machine Learning Applications for Bioenergy Conversion*, Green Energy and Technology, https://doi.org/10.1007/978-981-96-8086-3_3

forming a robust network within the cell wall that contributes to the biomass's resistance to degradation, a property known as "recalcitrance." This recalcitrance serves to protect the plant cell wall from being broken down by pathogens, microorganisms and enzymes. In this structure, cellulose acts as the main framework, forming strong microfibers, while hemicellulose and lignin fill in the gaps, connecting through hydrogen and covalent bonds (Wang et al. 2017). Cellulose provides mechanical strength and rigidity to plants, making CRMs inherently robust and fibrous.

Cellulose-rich materials are organic materials that contain a high proportion of cellulose, a polysaccharide that is the primary structural component of plant cell walls. Cellulose is the most abundant organic polymer on Earth and plays a crucial role in maintaining the rigidity and strength of plant structures. CRMs typically contain 40–60% cellulose, though some sources can have even higher concentrations. As plant-derived materials, CRMs are renewable resources and are biodegradable, making them environmentally friendly alternatives to fossil-based materials.

The following are examples of CRM's:

- Wood: One of the most common sources of cellulose, used extensively in paper, pulp, and construction industries.
- Agricultural Residues: Crop leftovers such as straw (wheat, rice, barley), corn stover, sugarcane bagasse, and cotton stalks.
- Grasses and Non-woody Plants: Switchgrass, miscanthus, bamboo, and hemp.
- Industrial Byproducts: Sawdust, wood chips, and other residues from timber processing.
- Dedicated Energy Crops: Plants like willow and poplar, grown specifically for biomass production.
- Algae and Seaweed: Some species contain significant amounts of cellulose and are being explored for biofuel and bioplastic production.

CRM's are used in many applications, including the following:

- Bioenergy Production: CRMs are a key feedstock for producing biofuels like ethanol (via enzymatic hydrolysis and fermentation) and biogas (through anaerobic digestion).
- Bioplastics and Biocomposites: Cellulose can be processed into bioplastics, films, and composites, offering sustainable alternatives to petroleum-based plastics.
- Paper and Pulp Industry: CRMs are the primary raw material for paper, cardboard, and other cellulose-based products.
- Textiles: Cellulose fibers are used to produce fabrics like rayon, lyocell, and modal.
- Chemical Feedstock: CRMs can be broken down into sugars and other chemicals for use in the production of bio-based chemicals and materials.
- Food and Pharmaceuticals: Cellulose derivatives like microcrystalline cellulose (MCC) and carboxymethyl cellulose (CMC) are used as stabilizers, thickeners, and dietary fibers.

There are several challenges in utilizing CRMs. As previously explained the recalcitrance of cellulose makes it resistant to degradation, necessitating efficient pretreatment and enzymatic processes. The composition of CRMs can vary widely depending on the source, which can affect processing efficiency and end-product quality. While CRMs are renewable, their large-scale use must be managed to avoid competition with food production and ensure sustainable land use practices.

One of the biggest challenges is lignin. The lignin component binds cellulose fibers and hinders the breakdown of cellulose. Lignin plays a crucial role as a barrier to breaking down lignocellulosic biomass for conversion into biofuels or biochemicals. By removing lignin—or partially removing hemicellulose along with it—the surface area of cellulose increases significantly. To remove or modify lignin requires pretreatment (e.g., chemical, thermal, or enzymatic). Pretreatments include various reactions such as hydrolysis, bio-digestion, and thermochemical processes. Ionic liquid solvents are particularly effective in pretreating lignocellulosic biomass by extracting lignin, resulting in a cellulose-rich material (CRM). This CRM can then be transformed into high-quality biofuels or chemicals through methods like pyrolysis (Khuenkaeo et al. 2020), torrefaction (Onsree and Tippayawong 2021), and fermentation (Dotsenko et al. 2018).

Ionic liquids (IL) are salts that remain in a liquid state at relatively low temperatures (typically below 100 °C). They are composed entirely of ions—a positively charged cation and a negatively charged anion, (Phromphithak et al. 2020). They are stable at high temperatures, which is useful for biomass pretreatment. Some key features of ionic liquids include:

- Non-flammability
- Low or negligible vapor pressure
- Chemical and thermal stability (Troter et al. 2016)
- Tunability: The properties of ILs can be tailored by selecting different cation-anion pairs, allowing them to dissolve specific components of biomass (e.g., cellulose, hemicellulose, or lignin).

These characteristics make ionic liquids a promising choice for effectively treating lignocellulosic biomass and enhancing its conversion into valuable products. ILs disrupt the hydrogen bonding network in cellulose, making it more accessible for enzymatic hydrolysis. They can also solubilize lignin and hemicellulose, separating them from cellulose and improving the efficiency of downstream processes. Ionic liquid solvents (ILSs) have gained considerable attention for their ability to reduce the recalcitrance of biomass. They can be classified into three main categories:

1. Ionic Liquids (ILs)
2. Deep Eutectic Solvents (DESs)
3. IL-Based Solvent Mixtures.

Recent studies have investigated the effectiveness of mixing ILs with organic solvents, revealing that these combinations can significantly enhance performance. The addition of organic solvents facilitates the dissociation of IL cations and anions, which in turn improves the solubility of lignocellulose, making biomass processing more efficient (Zhang et al. 2016b). These co-ionic liquid solvents are mixtures of ionic liquids with other solvents, such as water, organic solvents, or even other

Table 3.1 Key differences between ILs and Co-ILs

Feature	Ionic liquids (ILs)	Co-Ionic liquid solvents (Co-ILs)
Composition	Pure ionic liquids (cation + anion)	Mixtures of ILs with other solvents (e.g., water, organic solvents)
Cost	Often expensive due to high purity	More cost-effective due to dilution with co-solvents
Viscosity	Can be high, making handling difficult	Lower viscosity due to the presence of co-solvents
Targeted pretreatment	General dissolution of biomass components	More selective and efficient for specific components (e.g., lignin or hemicellulose)
Environmental impact	Low volatility, but some ILs may be toxic	Reduced toxicity and environmental impact due to dilution

ionic liquids. These mixtures are designed to enhance the pretreatment process by combining the unique properties of ILs with the advantages of the co-solvent. The addition of a co-solvent can improve the solubility of specific biomass components or reduce the viscosity of the IL, making it easier to handle. Co-ILs often require less pure IL, reducing overall costs. Co-ILs can target specific biomass components more effectively than pure ILs. For example, water as a co-solvent can help hydrolyze hemicellulose, while organic solvents can enhance lignin extraction. They can also reduce the energy requirements for pretreatment by lowering the operating temperature or shortening the processing time (Table 3.1).

Another innovative approach involves deep eutectic solvents (DESs), which are formed by combining Lewis or Brønsted acids and bases. These mixtures have melting points that are considerably lower than those of their individual components (Chen and Mu 2019). By modifying the structure or ratio of the constituent compounds, DESs can be tailored for various applications in areas such as catalysis, separation, and electrochemistry. The strong hydrogen bonds formed between hydrogen bond donors (HBD) and hydrogen bond acceptors (HBA) in DESs help to reduce biomass recalcitrance (Yu et al. 2018).

Interest in utilizing ILs for biomass pretreatment is growing (Zuin and Ramin 2018; Satlewal et al. 2018). In this chapter, we will apply machine learning (ML) techniques to unravel the complex relationships involved in biomass pretreatment using ILs. This task is well-suited for algorithms capable of identifying patterns within large datasets. ML will not only facilitate predictions but also evaluate the significance of various input features, providing valuable insights for experimental designs. The most effective ML algorithms for this analysis include support vector machines (SVM), random forest (RF), and gradient boosting (GB), all of which demonstrate reliable performance on datasets ranging from 17 to 800 entries with low root mean square error (RMSE) values.

Describing the characteristics of ILs presents a unique challenge. For example, Xu et al. (2020) explored the interrelationships among key variables in DES that influence lignocellulosic biomass pretreatment. Techniques such as principal component analysis (PCA) and partial least squares (PLS) were employed to examine the connections between the physical and chemical parameters of DESs, particularly those related to hydrogen bonding. In this chapter, ML models are used to predict and assess feature importance concerning solid recovery and cellulose content in CRMs. Using a dataset with 23 features—including biomass characteristics, ILs treatment conditions, IL properties, and catalyst levels—3 ML models were tested to determine which provides the most accurate predictions. Additionally, the predictive impact of different input features were evaluated by experimenting with the feature sets. Finally, the influence of highly significant features are visualized through one- and two-way partial dependence plots (PDPs), offering a clearer understanding of the intricate, multidimensional relationships within the ILs-based biomass pretreatment system.

3.2 Data Collection and Pre-processing

The datasets examined in this chapter focus on processes that quantify the CEF and SR from ionic solvent treatment of cellulose-rich materials. These include 23 related variables or features, all derived from ionic liquid solvent (ILS) pretreatment using conventional heating methods. This information was gathered through an extensive literature review, utilizing major scientific databases such as ScienceDirect, ISI Web of Sciences, the American Chemical Society, and the Royal Society of Chemistry. The review encompassed hundreds of peer-reviewed articles published between 2010 and 2020. To pinpoint the key factors effecting ILS pretreatment of biomass, the data was categorized into five main groups:

1. Lignocellulosic biomass characteristics
2. Pretreatment conditions
3. Main ILS identity
4. Co-ILS identity
5. Catalyst loading.

Each variable received a shorthand code to facilitate use in machine learning models, as detailed in Table 3.2. The primary ionic liquid solvent component was identified, while the co-ionic liquid solvent could be either an organic solvent or, in the case of deep eutectic solvents, a hydrogen bond donor or acceptor (HBD/HBA). This literature review resulted in approximately 850 data points. After excluding entries with incomplete values for the 23 input variables, the final dataset consisted of about 520 data points, representing around 45 biomass types and 80 solvent types.

In studies, the definition of cellulose content in cellulose-rich materials can vary. To simplify this chapter, the cellulose content in these materials was calculated as follows:

Table 3.2 Variables/features inputs for machine learning algorithms, from Phromphithak et al. (2021) (with permission)

Variable	Full name	Abbreviation	Value	Mean	SD
Input (Feature)					
Biomass property	Cellulose (% w/w)	Ce	16.8–61.1	34.8	8.7
	Hemicellulose (% w/w)	Hm	5.2–47.6	29.6	7.1
	Lignin (% w/w)	Lg	5.14–45.4	22.5	6.7
Treatment condition	Temperature (°C)	T	30–170	110	28
	Reaction time (h)	Rt	0.1–70	7.6	7.8
	Biomass loading (% w/w)	BL	0.4–50	10.0	8.9
Main ILS identity	The mole fraction of main solvent (–)	MFS1	0–1	0.6	0.4
	Molecular weight (g/mol)	MW1	0–277.4	157.7	47.5
	Carbon number (–)	Cn1	0–14	6.3	2.3
	Number of OH groups (–)	OH1	0–5	1.0	0.7
	Number of Amine groups (–)	NR1	0–5	1.5	0.8
	Number of Carbonyl groups (–)	CO1	0–1	0.3	0.5
	Number of Sulfur oxide groups (–)	SO1	0–1	0.1	0.2
	Chloride number (–)	CL1	0–1	0.5	0.5
Co-ILS identity	The mole fraction of co-solvent (–)	MFS2	0–1	0.4	0.4
	Molecular weight (g/mol)	MW2	0–198.3	42.7	41.8
	Carbon number (–)	CN2	0–10	1.4	1.6
	Number of OH groups (–)	OH2	0–5	1.2	1.2
	Number of Amine groups (–)	NR2	0–3	0.1	0.4
	Number of Carbonyl groups (–)	CO2	0–2	0.2	0.4
	Number of Sulfur oxide groups (–)	SO2	0–1	0.01	0.2
	Chloride number (–)	CL2	0–1	0.01	0.1
Catalyst	Catalyst dosage (% w/w)	Catalyst	0–4.33	0.1	0.4
Output (Target)					
CRM property	Cellulose enrichment factor (–)	CEF	0.2–5.2	2.2	1.0
	Solid recovery (% w/w)	SR	6–100	70.1	2.2

$$CEF = \frac{C_{e_C}}{C_{e_B}} \tag{3.1}$$

where CEF is the cellulose enrichment factor, C_{e_C} is the percent weight of cellulose in CRM, and C_{e_B} is the percent weight of cellulose in the original biomass.

Standardization and normalization of this dataset were also necessary for ML algorithms, Yeo-Johnson's non-linear transformation (Yeo and Johnson 2000) was used to standardize this dataset as described in Sect. 2.7.

3.3 Machine Learning Model Selection

In this chapter, the Random Forest (RF) model had the highest prediction accuracy based on metrics such as RMSE and R^2. When predicting the CEF, RF achieved an RMSE of 0.22 and an R^2 value close to 0.94. For solid recovery, RF recorded an RMSE of approximately 0.077 and an R^2 value exceeding 0.82. These results surpassed those of both support vector machines (SVM) and gradient boosting (GB) models, which had RMSE values greater than 0.25 and R^2 values below 0.80 for solid recovery, as shown in Fig. 3.1. The best hyperparameters for each model are listed in Table 3.3.

Random Forest regression is particularly effective with large datasets because it combines predictions from multiple decision tree algorithms, as discussed in Sect. 2.2. Each decision tree within the model is developed independently, enhancing the model's ability to generalize and predict accurately (Benimam et al. 2020). The main hyperparameters for RF include the number of trees, the minimum number of samples required to split an internal node, and the maximum number of leaf nodes in a best-first growth strategy. The optimal values for these hyperparameters (see Sect. 2.1.2) can be found in Table 3.3, while other settings were maintained at their default values in the scikit-learn library.

3.4 Feature Evaluation

To assess how different input features influence the prediction of the CEF and SR, this section explores four distinct datasets:

1. Dataset without all ionic liquid solvent features
2. Dataset without the main ionic liquid solvent features
3. Dataset without the co-ionic liquid solvent features
4. Full-feature dataset.

Each of these datasets was fine-tuned for its specific set of hyperparameters through a thorough grid search with tenfold cross-validation. Tenfold cross-validation is a widely used technique in machine learning for evaluating the performance of a model. It is a type of k-fold cross-validation, where the dataset is divided into k subsets (or "folds"), and the model is trained and validated k times. In the case of tenfold cross-validation, $k = 10$. The model is trained and evaluated 10 times. In each iteration, 1 fold is used as the validation set (or test set). The remaining 9 folds are combined and used as the training set. The model is trained on the 9 training folds and evaluated on the 1 validation fold. After all 10 iterations, the performance metrics from each fold are averaged to produce a single, robust estimate of the model's performance.

The base input features included characteristics of lignocellulosic biomass, pretreatment conditions, and catalyst information. Since Random Forest (RF) can rank feature importance during its decision tree construction, we utilized these rankings to understand how each feature impacts biomass pretreatment with ionic liquid solvents.

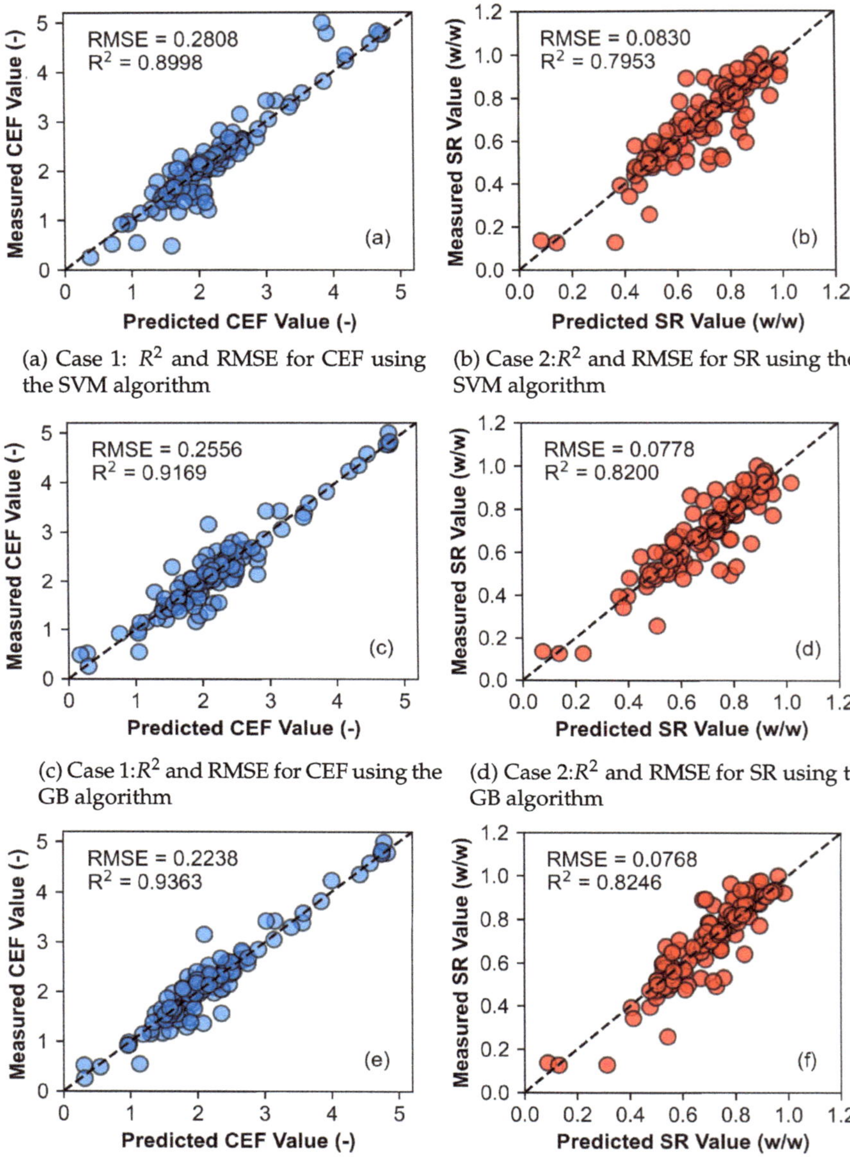

(a) Case 1: R^2 and RMSE for CEF using the SVM algorithm

(b) Case 2:R^2 and RMSE for SR using the SVM algorithm

(c) Case 1:R^2 and RMSE for CEF using the GB algorithm

(d) Case 2:R^2 and RMSE for SR using the GB algorithm

(e) Case 1:R^2 and RMSE for CEF using the RF algorithm

(f) Case 2:R^2 and RMSE for SR using the RF algorithm

Fig. 3.1 Machine learning model prediction accuracy via $RMSE$ and R^2 for CEF (**a**), (**c**), and (**e**), as well as for solid recovery (**b**), (**d**), and (**f**), from the three different algorithms of SVM in (**a**) and (**b**), GB in (**c**) and (**d**), and RF (**e**) and (**f**) from Phromphithak et al. (2021) (with permission)

Table 3.3 Optimized hyperparameters of the random forest, from Phromphithak et al. (2021) (with permission)

Algorithm	Hyperparameter	Optimized value
Random forest	'n_estimators'	310
	'max_depth'	None
	'max_leaf_nodes'	235
	'min_samples_split'	3
	'random_state'	30

As shown in Fig. 3.2, the model's prediction accuracy (measured by RMSE and R^2) varied based on the input features used. The dashed line in the figure indicates perfect alignment between predictions and actual test data. The model's accuracy changed significantly when certain features were added or removed. For instance, in

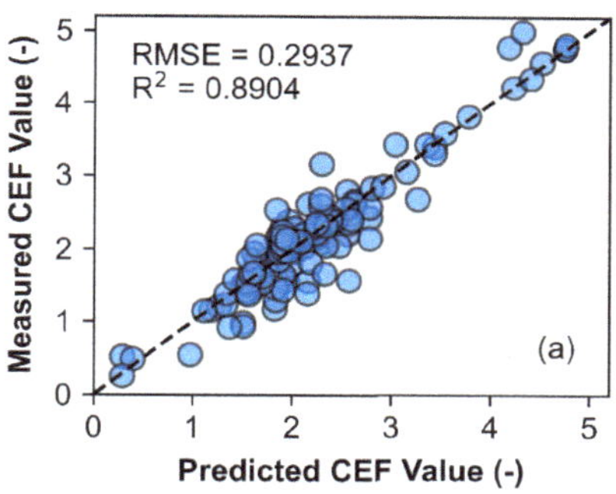

(a) Case 1: R^2 and RMSE for CEF excluding ionic liquid solvent features

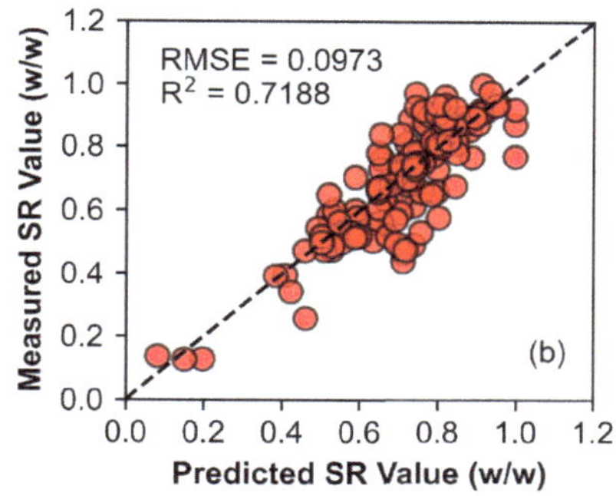

(b) Case 2:R^2 and RMSE for SR excluding ionic liquid solvent features

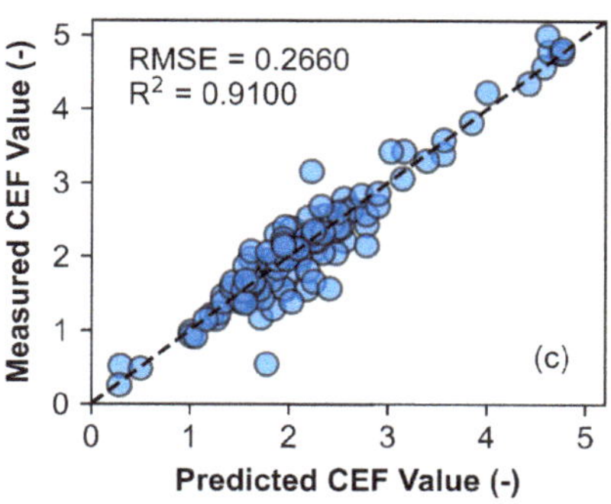

(c) Case 1:R^2 and RMSE for CEF excluding main ionic liquid solvent features

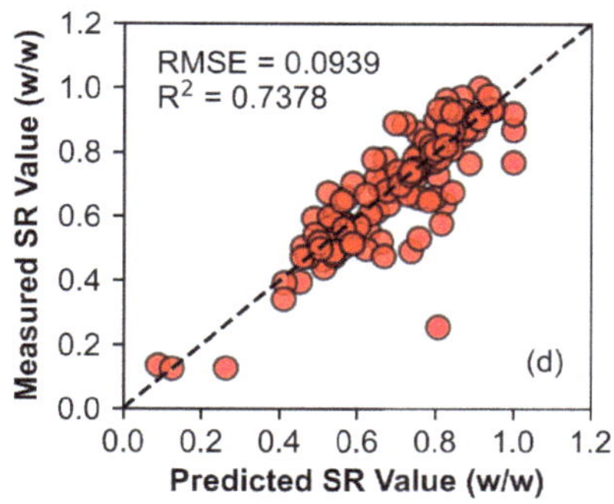

(d) Case 2:R^2 and RMSE for SR excluding main ionic liquid solvent features

Fig. 3.2 Machine learning model prediction accuracy via $RMSE$ and R^2 for CEF (**a**), (**c**), (**e**), and (**g**) as well as for SR (**b**), (**d**), (**f**), and (**h**) with four different feature datasets: (i) without any ionic liquid solvent features in (**a**) and (**b**), (ii) without the main ionic liquid solvent features in (**c**) and (**d**), (iii) without the co-ionic liquid solvent features in (**e**) and (**f**), and (iv) all the features in (**g**) and (**h**). The dash line represents positions that the predicted values are exactly equal to the test values, from Phromphithak et al. (2021) (with permission)

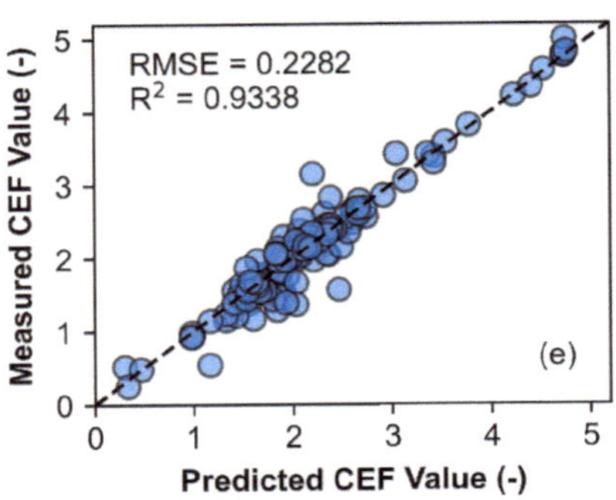

(e) Case 1:R^2 and RMSE for CEF excluding co- ionic liquid solvent features

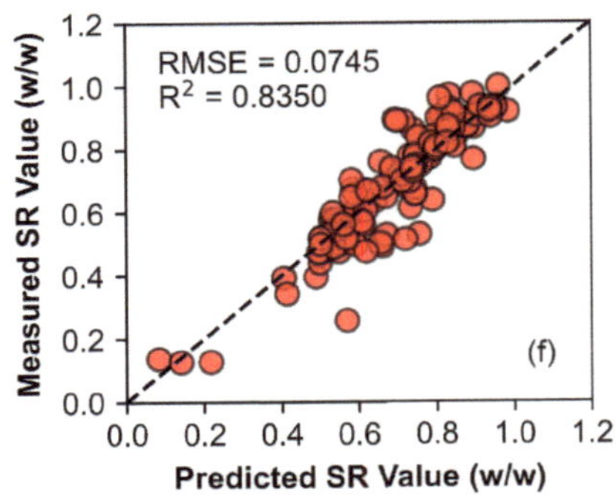

(f) Case 2:R^2 and RMSE for SR excluding co- ionic liquid solvent features

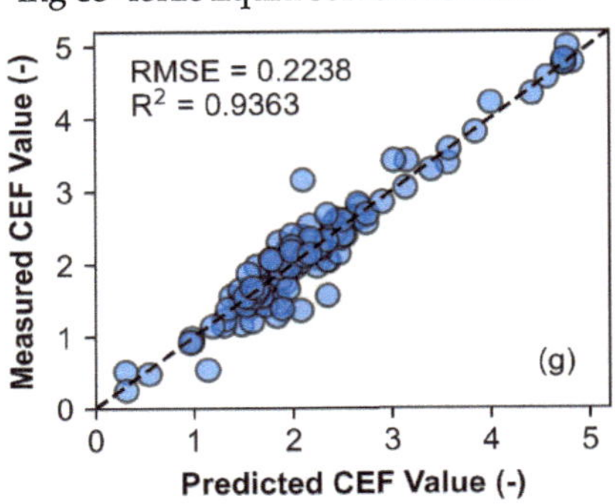

(g) Case 2:R^2 and RMSE for CEF excluding all solvent features

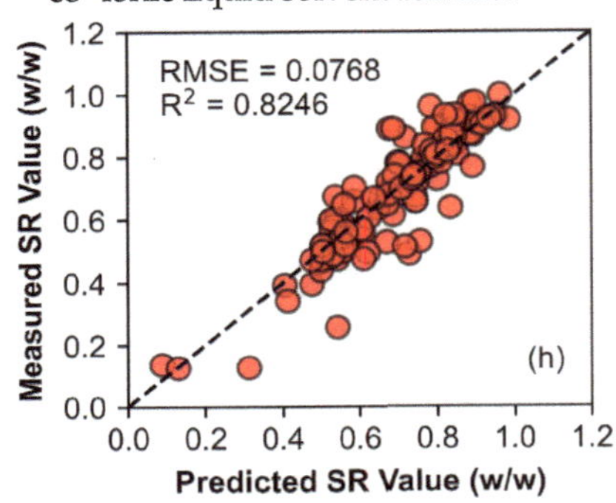

(h) Case 2:R^2 and RMSE for SR excluding all solvent features

Fig. 3.2 (continued)

Fig. 3.2a and b, which represent the base case using only raw biomass, pretreatment conditions, and catalyst features, RMSE was highest, and R^2 was lowest (0.29 and 0.89 for cellulose enrichment factor; 0.10 and 0.72 for solid recovery). However, incorporating features related to ionic liquid solvents greatly enhanced accuracy. For example, including co-ionic liquid solvent features, as depicted in Fig. 3.2e, led to a lower RMSE of 0.23 and an R^2 value as high as 0.93 for the cellulose enrichment factor. With the full-feature dataset (Fig. 3.2g and h), the model achieved its best performance for predicting the cellulose enrichment factor, showing an RMSE of 0.22 and an R^2 of 0.94. For solid recovery, the RMSE and R^2 values were 0.08 and 0.82, respectively. The highest accuracy for solid recovery occurred when using the base dataset combined with main ionic liquid solvent features, resulting in an RMSE of 0.07 and an R^2 of 0.84 (see Fig. 3.2e and f). This indicates that adding more features does not always enhance model accuracy; rather, the machine learning model benefits most from features that are closely related to the target variables. These findings are consistent with those of Pathy et al. (2020), who applied machine learning models to predict biochar yields from algal pyrolysis. They identified 13 important features based on raw biomass characteristics (derived from elemental and proximate analyses) and pyrolysis conditions, such as heating rate, reaction time, and pyrolysis temperature. Their model accurately predicted biochar yields with an R^2 of 0.81 using just four features, although accuracy slightly decreased (R^2 = 0.74) when employing the full-feature set. Similarly, in this study, the machine learning model achieved high predictive performance for both CEF and SR, comparable to previous machine learning research.

3.4.1 Feature Importance and Interpretation

The relative importance of each feature in predicting CEF and SR is shown in Fig. 3.3. The most influential features came from two groups: lignocellulosic biomass characteristics and pretreatment conditions. Notably, the cellulose content in raw biomass (Ce feature) was the most impactful feature for predicting CEF, while pretreatment temperature (T feature) was key for SR.

The Ce, Hm, and Lg features accounted for nearly 60% of the contribution to the CEF and around 30% for SR. The lignocellulosic composition of biomass impacts its resistance to breakdown; for example, biomass with lower lignin or hemicellulose

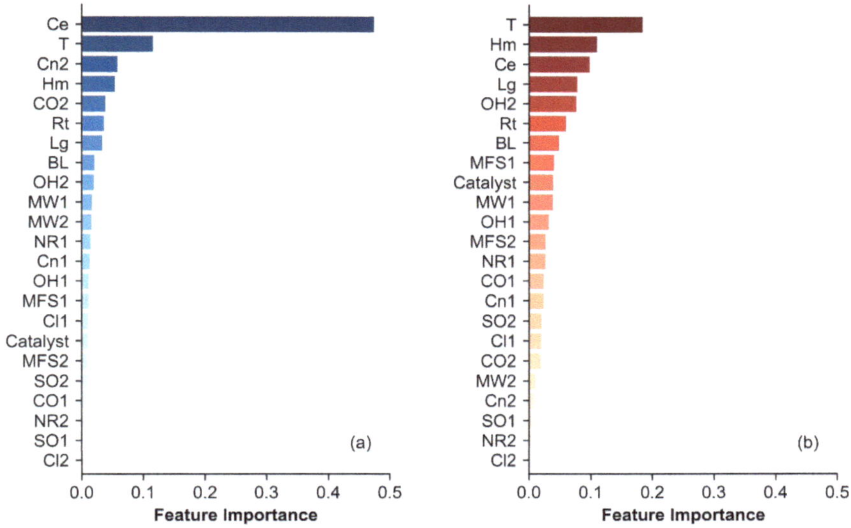

(a) Case 1: Feature importance for cellulose enrichment factor

(b) Case 2: Feature importance for solid recovery

Fig. 3.3 Feature importance for cellulose enrichment factor (**a**) and solid recovery (**b**), from Phromphithak et al. (2021) (with permission) content can be decomposed more easily, allowing ionic liquid solvents to access and reduce cellulose crystallinity. This results in a solid rich in cellulose (Halder et al. 2019). Similarly, Procentese et al. (2018), used deep eutectic solvents for pretreating agricultural waste, found that even with uniform conditions, different biomass compositions could yield CRM's with varying properties. Other biomass characteristics, such as cellulose crystallinity, ash content, and moisture content, may also serve as influential input features.

For pretreatment conditions (T, Rt, and BL features), the T feature emerged as the second most critical factor for predicting the CEF. Collectively, features from pretreatment conditions contributed about 20% to this. For SR, the T feature also had the highest importance, and the combined impact of T, Rt, and BL reached 30%, equaling the contribution from biomass characteristics. Cox and Ekerdt (2012) studied how ionic liquid solvents depolymerize lignin into valuable phenolic compounds, finding that optimal reaction temperature and time were crucial. Adjusting these parameters significantly influenced lignin depolymerization by breaking alkyl-aryl ether bonds via hydrolysis.

While features from IL's and co-IL's were lower in the importance ranking, they still contributed meaningfully, accounting for over 35% for SR and 18% for the CEF. Ionic liquid solvent features help illustrate how the chemical structure of ionic liquids affects biomass pretreatment, potentially guiding new solvent designs. Due to their structural complexity, ionic liquids encompass a wide range of potential applications, making them challenging to model comprehensively. In this study, ionic liquid solvents were simplified by categorizing them based on their chemical molecular structure relevant to biomass pretreatment. The variables associated with ionic liquid solvents in this context include:

1. Functional groups (e.g., carbonyl, hydroxyl, amine, sulfur oxide, chloride) that create intermolecular forces, helping to break down adhesion in lignocellulosic biomass.
2. Molecular weight and carbon number, which may influence surface reactions through steric effects.
3. Mole fraction, representing properties like viscosity, solubility, and equilibrium.

For CEF, the carbon number (Cn2 feature) for IL features and the carbonyl groups ($CO2$ feature) in co-IL's were the most impactful. For SR, hydroxyl groups ($OH2$ feature) in co-IL's and the mole fraction (MFS1 feature) in IL's were most significant. These findings underscore that both main and co-ionic liquid solvent features are important for accurate predictions of CEF and SR. For example, excluding co-ionic liquid solvent features slightly reduced R^2 for SR (as shown in Fig. 3.2f and h).

This approach to IL solvent identification aligns with Yalcin et al. (2019), who used ML models to create structure-property models for protic ionic liquids. Their dataset, based on fundamental chemical structures, found that factors like alkyl chain length, hydroxyl groups of cations, anion types, nonstoichiometry, and water presence significantly impacted the resulting properties, enhancing the application potential of ionic liquid mixtures. Other properties, such as Kamlet-Taft solvent parameters and σ -profile, could further refine ionic liquid property prediction methods (Benimam et al. 2020; Kang et al. 2017; Xu et al. 2020). However, gathering these parameters for all ionic liquids remains challenging.

3.4.2 *Partial Dependence Plots*

The one-way partial dependence plots (PDPs) for the operating conditions of biomass pretreatment with ionic liquids (ILs) are shown in Fig. 3.4. In Fig. 3.4a for CEF and (b) for (SR), the temperature (T) feature had a similar effect on both. As the temperature increased, both CEF and SR decreased linearly. This is because higher temperatures led to greater extraction of lignin, which reduced SR. At the same time, the solubility of cellulose increased, meaning more cellulose dissolved into the ILs, resulting in a lower CEF value.

When examining the reaction time (Rt) feature in Fig. 3.4c and d, we observed different trends for CEF and SR. In Fig. 3.4d, SR decreased as the reaction time increased. However, in Fig. 3.4c, CEF initially rose with reaction time, reached a peak, and then declined as the reaction time continued to increase. The initial increase in CEF aligns with findings from Zhao et al. (2018), who suggested that longer reaction times could enhance cellulose content. On the other hand, the subsequent decline in CEF with prolonged reaction time is consistent with Dotsenko et al. (2018), who reported that excessive reaction times could negatively impact cellulose content.

The impact of biomass loading (BL) on CEF and SR is illustrated in Fig. 3.4e and f, respectively. The relationship follows an L-shaped curve, indicating that lower biomass loadings yielded better results. As biomass loading increased, the surface area available for interaction between the biomass and ILs decreased, leading to reduced removal of lignin and xylan (Chen and Mu 2019). Further increases in

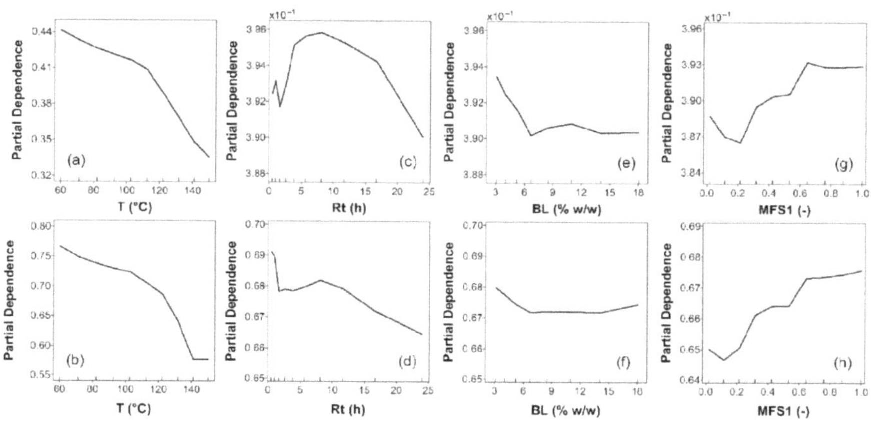

Fig. 3.4 One-way PDPs of the features from pretreatment conditions for CEF (**a**), (**c**), (**e**), and (**g**) as well as for SR (**b**), (**d**), (**f**), and (**h**), from Phromphithak et al. (2021) (with permission)

biomass loading did not significantly improve lignin extraction or CEF but limited the ability of ILs to access cellulose. Lignin acts as a binding agent that holds lignocellulosic material together. When a portion of lignin is removed, the crystalline structure of cellulose begins to break down. These findings are supported by Wu et al. (2011), who studied biomass pretreatment with ILs at room temperature for high biomass loadings. They found that lignin extraction efficiency decreased at biomass loadings above 9.1% w/w. At loadings greater than 33% w/w, the crystalline structure of cellulose began to resemble that of untreated biomass.

The effect of the mole fraction between the main IL and co-IL is shown in Fig. 3.4g and h, where the MFS1 feature represents one of these components. The MFS1 feature had a linear effect on both CEF and SR. The CRM improved when the mole fraction of the main IL was higher than that of the co-IL. The IL-based solvent system must strike a balance between breaking down the recalcitrance of lignocellulosic biomass and dissolving the resulting components (Zhang et al. 2016b). Similarly, in deep eutectic solvents (DESs), the mole ratio of hydrogen bond acceptors (HBA) and hydrogen bond donors (HBD) must be optimized to form hydrogen bonds that effectively break down the recalcitrance of lignocellulosic biomass (Zhang et al. 2016a, b).

To explore how pairs of features interact and influence the results, two-way partial dependence plots (PDPs) were constructed. Similar to the one-way PDPs, the focus was on features that had the most significant impact on the outputs. Figure 3.5 shows the interactions between temperature (T) and reaction time (Rt), as well as the mole fraction of the main ionic liquid solvent (ILS) and characteristics of the lignocellulosic biomass.

In Fig. 3.5a and b, the interaction between T and Rt had a relatively minor influence on both CEF and SR compared to their individual effects. Temperature had a stronger influence on CEF and SR than reaction time. However, increasing the reaction time

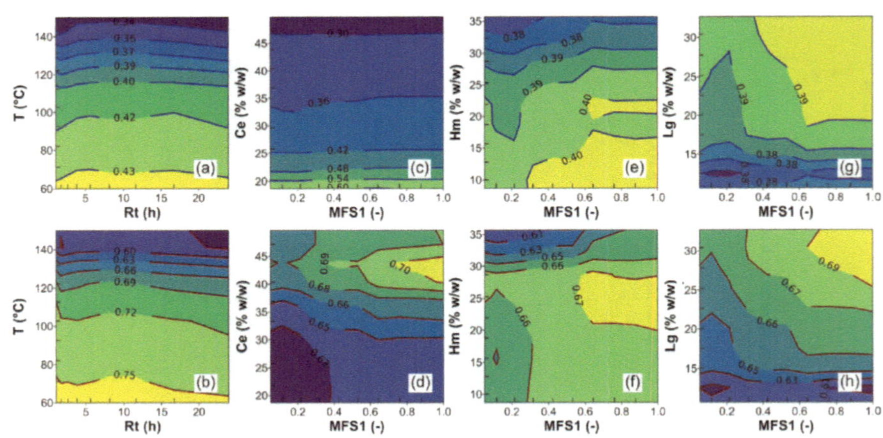

Fig. 3.5 Two-way PDPs of the features from pretreatment conditions for CEF (**a**), (**c**), (**e**), and (**g**) as well as for SR (**b**), (**d**), (**f**), and (**h**), from Phromphithak et al. (2021) (with permission)

could still help improve lignin extraction and enhance the solubility of lignocellulosic biomass fragments.

Figure 3.5c to h illustrates the role of the mole fraction, represented by the MFS1 feature, in CEF and SR during biomass pretreatment with ILs. In Fig. 3.5c, the interaction between the MFS1 feature and cellulose content (Ce) did not significantly affect CEF. This suggests that the combination of mole fraction solvents and cellulose content in raw biomass did not lead to cellulose depolymerization. In contrast, Fig. 3.5d shows that varying the mole fraction of solvents had a noticeable effect on SR. This is primarily because ILs break down hemicellulose and lignin, which are key components influencing solids recovery.

In Fig. 3.5e to h, the interaction between the mole fraction solvents and the hemicellulose and lignin content in raw biomass showed clear relationships with both CEF and SR. For biomass with unknown properties, the results from these two-way PDPs can serve as a guideline to determine the optimal range of mole fraction solvents for pretreatment with ILs, based on the composition of the lignocellulosic biomass.

3.5 Conclusion

The Random Forest model was effectively used to predict both the CEF and SR in cellulose-rich materials obtained through biomass pretreatment with ionic liquid solvents. This was achieved using datasets with 23 features. The features with the greatest impact, contributing approximately 80% to the CEF and 60% to SR, were identified as the cellulose content and pretreatment temperature.

References

Benimam H, Moussa CS, Hentabli M, Hanini S, Laidi M (2020) Dragonfly-support vector machine for regression modeling of the activity coefficient at infinite dilution of solutes in imidazolium ionic liquids using σ-profile descriptors. J Chem Eng Data 65(6):3161–3172. ISSN 0021-9568. https://doi.org/10.1021/acs.jced.0c00168

Chen Y, Mu T (2019) Application of deep eutectic solvents in biomass pretreatment and conversion. Green Energy & Environ 4(2):95–115. ISSN 2468-0257. https://doi.org/10.1016/j.gee.2019.01.012. https://www.sciencedirect.com/science/article/pii/S2468025718301419. Special Issue: Ionic liquids in energy and environment

Cox BJ, Ekerdt JG (2012) Depolymerization of oak wood lignin under mild conditions using the acidic ionic liquid 1-h-3-methylimidazolium chloride as both solvent and catalyst. Bioresour Technol 118:584–588. ISSN 0960-8524. https://doi.org/10.1016/j.biortech.2012.05.012. https://www.sciencedirect.com/science/article/pii/S0960852412007559

Dotsenko AS, Dotsenko GS, Senko OV, Stepanov NA, Lyagin IV, Efremenko EN, Gusakov AV, Zorov IN, Rubtsova EA (2018) Complex effect of lignocellulosic biomass pretreatment with 1-butyl-3-methylimidazolium chloride ionic liquid on various aspects of ethanol and fumaric acid production by immobilized cells within SSF. Bioresour Technol 250: 429–438. ISSN 0960-8524. https://doi.org/10.1016/j.biortech.2017.11.064. https://www.sciencedirect.com/science/article/pii/S0960852417320680

Halder P, Kundu S, Patel S, Setiawan A, Atkin R, Parthasarthy R, Paz-Ferreiro J, Surapaneni A, Shah K (2019) Progress on the pre-treatment of lignocellulosic biomass employing ionic liquids. Renew Sustain Energy Rev 105:268–292. ISSN 1364-0321. https://doi.org/10.1016/j.rser.2019.01.052. https://www.sciencedirect.com/science/article/pii/S1364032119300784

Kang X, Zhao Z, Qian J, Muhammad Afzal R (2017) Predicting the viscosity of ionic liquids by the ELM intelligence algorithm. Ind Eng Chem Res 56(39):11344–11351. ISSN 0888-5885. https://doi.org/10.1021/acs.iecr.7b02722

Khuenkaeo N, MacQueen B, Onsree T, Daiya S, Tippayawong N, Lauterbach J (2020) Bio-oils from vacuum ablative pyrolysis of torrefied tobacco residues. RSC Adv 10:34986–34995. https://doi.org/10.1039/D0RA06014C

Onsree T, Tippayawong N (2021) Machine learning application to predict yields of solid products from biomass torrefaction. Renew Energy 167:425–432. ISSN 0960-1481. https://doi.org/10.1016/j.renene.2020.11.099. https://www.sciencedirect.com/science/article/pii/S0960148120318516

Pathy A, Meher S (2020) Predicting algal biochar yield using extreme gradient boosting (XGB) algorithm of machine learning methods. Algal Res 50:102006. ISSN 2211-9264. https://doi.org/10.1016/j.algal.2020.102006. https://www.sciencedirect.com/science/article/pii/S2211926420303684

Phromphithak S, Meepowpan P, Shimpalee S, Tippayawong N (2020) Transesterification of palm oil into biodiesel using ChOH ionic liquid in a microwave heated continuous flow reactor. Renew Energy 154:925–936. ISSN 0960-1481. https://doi.org/10.1016/j.renene.2020.03.080. https://www.sciencedirect.com/science/article/pii/S0960148120304055

Phromphithak S, Onsree T, Tippayawong N (2021) Machine learning prediction of cellulose-rich materials from biomass pretreatment with ionic liquid solvents. Bioresour Technol 323:124642. ISSN 0960-8524. https://doi.org/10.1016/j.biortech.2020.124642. https://www.sciencedirect.com/science/article/pii/S0960852420319167

Procentese A, Raganati F, Olivieri G, Russo ME, Rehmann L, Marzocchella A (2018) Deep eutectic solvents pretreatment of agro-industrial food waste. Biotechnol Biofuels 11(1):37. ISSN 1754-6834. https://doi.org/10.1186/s13068-018-1034-y

Satlewal A, Agrawal R, Bhagia S, Sangoro J, Ragauskas AJ (2018) Natural deep eutectic solvents for lignocellulosic biomass pretreatment: recent developments, challenges and novel opportunities. Biotechnol Adv 36(8):2032–2050. ISSN 0734-9750. https://doi.org/10.1016/j.biotechadv.2018.08.009. https://www.sciencedirect.com/science/article/pii/S0734975018301472

Tang Q, Chen Y, Yang H, Liu M, Xiao H, Wu Z, Chen H, Naqvi SR (2020) Prediction of bio-oil yield and hydrogen contents based on machine learning method: effect of biomass compositions and pyrolysis conditions

Troter DZ, Todorovic ZB, Dokic-Stojanovic DR, Stamenkovic OS, Veljkovic VB (2016) Application of ionic liquids and deep eutectic solvents in biodiesel production: a review. Renew Sustain Energy Rev 61:473–500. ISSN 1364-0321. https://doi.org/10.1016/j.rser.2016.04.011. https://www.sciencedirect.com/science/article/pii/S1364032116300284

Wang S, Dai G, Yang H, Luo Z (2017) Lignocellulosic biomass pyrolysis mechanism: a state-of-the-art review. Prog Energy Combust Sci 62:33–86. ISSN 0360-1285. https://doi.org/10.1016/j.pecs.2017.05.004. https://www.sciencedirect.com/science/article/pii/S0360128517300266

Wu H, Mora-Pale M, Miao J, Doherty TV, Linhardt RJ, Dordick JS (2011) Facile pretreatment of lignocellulosic biomass at high loadings in room temperature ionic liquids. Biotechnol Bioeng 108(12):2865–2875. https://doi.org/10.1002/bit.23266

Xu H, Kong Y, Peng J, Song X, Che X, Liu S, Tian W (2020a) Multivariate analysis of the process of deep eutectic solvent pretreatment of lignocellulosic biomass. Ind Crop Prod 150:112363. ISSN 0926-6690. https://doi.org/10.1016/j.indcrop.2020.112363. https://www.sciencedirect.com/science/article/pii/S092666902030279X

Xu X, Range J, Gygli G, Pleiss J (2020b) Analysis of thermophysical properties of deep eutectic solvents by data integration. J Chem Eng Data 65(3):1172–1179. ISSN 0021-9568. https://doi.org/10.1021/acs.jced.9b00555

Yalcin D, Le TC, Drummond CJ, Greaves TL (2019) Machine learning approaches for further developing the understanding of the property trends observed in protic ionic liquid containing solvents. J Phys Chem B 123:4085–4097

Yeo I-K, Johnson RA (2000) A new family of power transformations to improve normality or symmetry. Biometrika 87(4):954–959. ISSN 0006-3444. https://doi.org/10.1093/biomet/87.4.954

Yu Q, Zhang A, Wang W, Chen L, Bai R, Zhuang X, Wang Q, Wang Z, Yuan Z (2018) Deep eutectic solvents from hemicellulose-derived acids for the cellulosic ethanol refining of akebia herbal residues. Bioresour Technol 247:705–710. ISSN 0960-8524. https://doi.org/10.1016/j.biortech.2017.09.159. https://www.sciencedirect.com/science/article/pii/S0960852417317248

Zhang C-W, Xia S-Q, Ma P-S (2016a) Facile pretreatment of lignocellulosic biomass using deep eutectic solvents. Bioresour Technol 219:1–5. ISSN 0960-8524. https://doi.org/10.1016/j.biortech.2016.07.026. https://www.sciencedirect.com/science/article/pii/S0960852416309907

Zhang X, Zhao W, Li Y, Li C, Yuan Q, Cheng G (2016b) Synergistic effect of pretreatment with dimethyl sulfoxide and an ionic liquid on enzymatic digestibility of white poplar and pine. RSC Adv 6:62278–62285. https://doi.org/10.1039/C6RA14206K

Zhao Z, Chen X, Ali MF, Abdeltawab AA, Yakout SM, Yu G (2018) Pretreatment of wheat straw using basic ethanolamine-based deep eutectic solvents for improving enzymatic hydrolysis. Bioresour Technol 263:325–333. ISSN 0960-8524. https://doi.org/10.1016/j.biortech.2018.05.016. https://www.sciencedirect.com/science/article/pii/S0960852418306655

Zuin VG, Ramin LZ (2018) Green and sustainable separation of natural products from agro-industrial waste: challenges, potentialities, and perspectives on emerging approaches. Top Curr Chem 376(1):3. ISSN 2364-8961. https://doi.org/10.1007/s41061-017-0182-z

Chapter 4
Torrefaction

Abstract In this chapter, machine learning is employed to develop a model that predicts the yield of solid products from biomass torrefaction. The model uses input features that describe both the characteristics of the biomass and the conditions under which the torrefaction process occurs. Several machine learning algorithms were evaluated using tenfold cross-validation, and their settings were fine-tuned through a detailed grid search. Among the tested algorithms, the gradient tree boosting method performed the best, achieving a prediction accuracy with an R^2 value of nearly 0.90 and an average error of 0.07 w/w. Six features were identified as particularly important for the model's predictions. For the torrefaction conditions, these included the temperature, the residence time, and the oxygen concentration in the reacting gas. For the biomass properties, the key features were the levels of volatile matter, carbon content, and oxygen content. Interestingly, all features except carbon content were found to decrease the yield of torrefied biomass. In terms of overall influence, the properties of the biomass contributed approximately 30% to the solid yield, with volatile matter alone making up about one-third of this contribution. This approach highlights how machine learning can effectively predict torrefaction outcomes by combining process conditions and biomass characteristics, offering valuable insights for optimizing the process.

Keywords Biomass · Gradient tree boosting · Machine learning · Solid fuels · Torrefaction

4.1 Introduction

Over the past few decades, the extensive use of fossil fuels for electricity generation and fuel production has led to serious challenges, not least including climate change (Bukhtiyarova et al. 2017). In response, biomass has gained attention as a renewable resource with the potential to replace fossil fuels in producing valuable energy sources (Onsree and Tippayawong 2017; Liu and Han 2015). Several advanced methods have been developed to convert various types of biomass into fuel (Fan et al. 2020; Onsree et al. 2018a). However, using raw biomass on a large scale remains difficult compared

N. Tippayawong et al., *Machine Learning Applications for Bioenergy Conversion*, Green Energy and Technology, https://doi.org/10.1007/978-981-96-8086-3_4

to fossil fuels. Raw biomass often has low bulk density, high moisture content, and low energy content, which can hinder the efficiency of conversion processes (Cahyanti et al. 2020). As a result, developing effective pre-treatment methods is essential to improve biomass quality and make it more competitive with fossil fuels.

One of the most promising pre-treatment methods is torrefaction, which addresses many of the limitations of raw biomass. Torrefaction, also referred to as mild pyrolysis, involves heating biomass at temperatures between 200 and 300 °C in a controlled environment. This process removes moisture and breaks down certain components, such as hemicelluloses, releasing oxygen-rich volatile gases like CO and CO_2 (Tian et al. 2020). The primary output, called torrefied biomass, has a higher energy content and is easier to grind. However, the amount of torrefied biomass produced can vary widely depending on the properties of the original biomass and the conditions of the torrefaction process. For instance, under the same torrefaction conditions –20 min at 250 °C—pine yields more than 85% solid mass, while eucalyptus yields around 78% (Alvarez et al. 2018). This variability highlights the importance of understanding how different factors influence the torrefaction process to optimize its outcomes.

To predict the yield of torrefied biomass for specific types, earlier studies have relied on models based on torrefaction kinetics or empirical methods such as simple regression and response surface analysis (Chiou et al. 2015; Zhang et al. 2020; Singh et al. 2020). These models typically use only a small number of variables, such as torrefaction temperature and residence time, which limits their accuracy and ability to generalize to different scenarios. In contrast, machine learning (ML) techniques offer a more powerful approach by analyzing complex, multidimensional relationships in biomass torrefaction using larger datasets. For example, Zhu et al. (2019) used a random forest ML algorithm to predict the yield and carbon content of solid products from biomass pyrolysis based on raw biomass characteristics. Their study, which used a dataset with twelve features, found that properties like ash content and hemicellulose content had a significant impact on solid yield and carbon content. Similarly, Garcia Nieto et al. (2019) applied ML to estimate the calorific values of torrefied biomass using raw biomass properties and torrefaction conditions, highlighting the importance of features such as volatile matter and the atomic O/C ratio.

Despite these advances, few studies have explored the use of ML to predict the yield of torrefied biomass or to determine which features have the greatest influence on this outcome. The goal of this chapter is to develop ML models that can accurately predict the yield of solid products from biomass torrefaction. This study uses a dataset with fourteen features, including both biomass characteristics and torrefaction conditions. Several ML models were trained and tested, with tenfold cross-validation used to evaluate their performance and identify the best model based on prediction accuracy. Additionally, the importance and influence of each feature on the model's predictions were analyzed using three different datasets:

1. A full dataset containing all fourteen features: moisture content (MC), volatile matter, fixed carbon (FC), ash, carbon content (C), hydrogen content (H), nitrogen content (N), oxygen content (O), sample size, residence time, temperature, and the concentrations of N_2, CO_2, and O_2 in the reacting gas.

2. A reduced dataset that excludes moisture content (MC), fixed carbon (FC), oxygen content (O), and the concentration of N_2 in the reacting gas.
3. A second reduced dataset that excludes carbon content (C), hydrogen content (H), nitrogen content (N), and oxygen content (O).

By analyzing these datasets, this chapter aims to provide insights into the key factors influencing torrefied biomass yield and to demonstrate how ML can be used to optimize the torrefaction process.

4.2 Materials and Methods

A dataset containing information on the yield of torrefied biomass, referred to as the solid yield (SY), along with its associated features, was compiled from previous experiments and published studies, resulting in over 800 data points. The fourteen features or variables used in this study are detailed in Table 4.1, which includes the name, range of values, and standard deviation for each feature. The uncertainty in the data for each variable, whether from in-house experiments or literature sources, was within 5%. The biomass feedstocks examined in this study include agricultural residues, forestry residues, and energy crops, all categorized under the broader term of lignocellulosic biomass. Typically, these feedstocks are dried and reduced in size to ensure uniformity before their physical and chemical properties are analyzed and used in torrefaction processes. The key factors influencing the solid yield of torrefied

Table 4.1 Set of input features/variables, from Onsree and Tippayawong (2021) (with permission)

Feature	Name	Value	Standard deviation
Biomass property (on dry basis)			
Moisture content (% w/w)	MC	0–73.2	11.9
Volatile matter (% w/w)	VM	47.6–94.4	8.8
Fixed carbon (% w/w)	FC	5.0–32.7	6.1
Ash (% w/w)	Ash	0.2–20.2	3.9
Carbon content (% w/w)	C	32.5–61.4	4.6
Hydrogen content (% w/w)	H	4.8–8.7	0.7
Nitrogen content (% w/w)	N	0–3.9	0.8
Oxygen content (% w/w)	O	25.9–61.6	5.8
Sample size (mm)	SZ	0.1–20	4.7
Torrefaction condition			
Residence time (min)	Time	5–120	20.1
Temperature (°C)	T	200–340	32.9
CO_2 fraction in the reacting gas (v/v)	$CO2$	0–1	0.1
O_2 fraction in the reacting gas (v/v)	$O2$	0–0.21	0.05
N_2 fraction in the reacting gas (v/v)	$N2$	0–1	0.1

biomass can be divided into two main categories: biomass properties and torrefaction conditions.

Although each data point for solid yield included complete values for all variables, it was still necessary to standardize the dataset for certain machine learning (ML) algorithms. Many ML algorithms may perform poorly if some input features deviate significantly from a standard normal (Gaussian) distribution. In this study, the 14-feature dataset was standardized using Yeo-Johnson's non-linear transformation, as described in Sect. 2.7. This step ensures that the data is appropriately scaled and prepared for accurate analysis and modeling.

4.2.1 Machine Learning

The machine learning (ML) framework for this study was developed using Python's scikit-learn library (Pedregosa et al. 2011). A variety of ML algorithms were tested, including lasso regression, normal ridge regression, kernel ridge regression, decision trees, adaptive boosting, gradient tree boosting, random forests, extremely randomized trees, k-nearest neighbors, and support vector machines. Since each dataset has a unique structure, no single algorithm is universally optimal for all problems (Williams et al. 2020).

To evaluate the prediction performance of each model in an unbiased manner, k-fold cross-validation was employed. This method involves randomly shuffling the dataset and dividing it into k non-overlapping partitions, or "folds." The ML model is trained on k–1 folds (The training data, $\{X, Y\}$) and tested on the remaining fold (test data, $\{x, y\}$). This process is repeated k times, with each fold serving as the test set once. This ensures that every data point is used for both training and testing, providing a robust evaluation of the model's performance.

For each model, a full-factor grid search was used to optimize the hyperparameters. This approach systematically explores all possible combinations of hyperparameters, combined with nested k-fold cross-validation. In this study, $k = 10$ was chosen, as it strikes a balance between providing an unbiased estimate of model performance and ensuring that each training and test group is large enough to represent the overall dataset (Elmaz et al. 2020; Kuhn and Johnson 2013). The combination of hyperparameters that delivered the best performance was selected as the optimal set for each model. Once tuned, the model was retrained using its optimal settings and evaluated based on the regression coefficient (R^2) and root mean squared error (RMSE), as outlined in Sect. 2.7.

The top-performing algorithms, based on tenfold cross-validation, were kernel ridge regression (KRR) and gradient tree boosting (GTB). Both algorithms achieved similar results, with R^2 values exceeding 0.86, RMSE values around 0.09, and normalized RMSE (NRMSE) values below 0.1, on average. Kernel ridge regression combines the "kernel trick" with ridge regression (linear least squares with L2-norm regularization) to model non-linear relationships in the dataset's original space. For gradient tree boosting, the optimal settings included 550 predictors, a learning rate of 0.1, a minimum sample split of 10, and a minimum sample leaf of 10. These configurations allowed the models to achieve high accuracy and reliability in predicting the solid yield of torrefied biomass.

4.2.2 Feature Evaluation

To explore how different input features influenced the accuracy of the machine learning (ML) model predictions, the models were trained and tested using three distinct datasets: (1) a dataset containing all fourteen features (Case 1); (2) a dataset excluding the N_2 concentration, moisture content (MC), fixed carbon (FC), and oxygen content (O2) features (Case 2); and (3) a dataset excluding carbon content (C), hydrogen content (H2), nitrogen content (N2), and oxygen content (O2) features (Case 3). Additionally, the relative importance of each feature was evaluated using the permutation importance method.

The permutation importance method works by measuring how much the model's prediction accuracy changes when a specific feature is either removed or randomly shuffled, while keeping the target variable and all other features unchanged (Fisher et al. 2019). This process is carried out after the model has been trained. First, the model's prediction accuracy is calculated as a reference score, denoted as S, using a metric such as the R^2 score. Then, the feature of interest is shuffled, and the model's accuracy is recalculated. The difference between the reference score (S) and the new score after shuffling indicates the importance of that feature. A larger drop in accuracy suggests that the feature plays a more significant role in the model's predictions.

This approach provides a clear understanding of which features contribute most to the model's performance and helps identify which variables are critical for accurate predictions. By analyzing the results from the three datasets and applying the permutation importance method, this study offers valuable insights into the key factors influencing the solid yield of torrefied biomass and highlights the effectiveness of ML in optimizing the torrefaction process.

For feature x_j, the original dataset D_j is randomly shuffled to generate a corrupted version $\tilde{D}_{t,j}$ with repetition t. The score $S_{t,j}$ of the model is computed on the corrupted $\tilde{D}_{t,j}$, and the permutation importance P_j is defined as

$$P_j = S - \frac{1}{n}\sum_{[t=1]}^{n} S_{t,j} \tag{4.1}$$

where n is the number of repetitions. In this method, the relationship between a given feature and the target variable is disrupted. Therefore, if the model's accuracy score changes significantly after disrupting a specific feature, it indicates that the model relies heavily on that feature for making predictions. To further investigate how each important feature influences the yield of solid products in the torrefaction process, partial dependence plots (PDPs) were used. PDPs provide a visual representation of the relationship between each feature and the target variable, showing whether the relationship is linear, monotonic, or more complex.

Partial dependence plots work by isolating the effect of a single feature while averaging out the influence of all other features. This allows us to see how changes in a specific feature impact the predicted outcome, holding all other variables constant. For example, a PDP might reveal that increasing the torrefaction temperature leads to a steady decrease in solid yield, indicating a linear relationship. Alternatively, it could show a more complex pattern, such as a peak in solid yield at a specific temperature before declining.

By analyzing these plots, we can gain deeper insights into how individual features affect the torrefaction process and identify optimal conditions for maximizing solid yield. This approach not only enhances our understanding of the process but also helps in fine-tuning the model for more accurate predictions.

4.3 Results

As shown in the Pearson correlation matrix in Fig. 4.1, the yield of torrefied biomass exhibited a strong linear relationship with torrefaction temperature ($p < 0.01$) and residence time ($p < 0.01$). The p-values, calculated using a two-tailed t-distribution with $N - 2$ degrees of freedom, indicate the statistical significance of these linear correlations. The correlation coefficients were –0.61 for temperature and –0.35 for residence time. These negative values indicate that as temperature and residence time increased, the yield of solid products from torrefaction decreased. This trend is consistent with findings from previous experimental studies (Mohd Faizal et al. 2018; Cardona et al. 2019; Kanwal et al. 2019; Khuenkaeo et al. 2020).

In contrast, other features showed only weak correlations with the solid yield, with absolute correlation coefficients below 0.15. However, some features were strongly correlated with each other. For example, within the biomass properties, carbon content and oxygen content had a high correlation of –0.91 ($p < 0.05$). Among the torrefaction conditions, the N_2 fraction in the reacting gas was strongly correlated with the CO_2 and O_2 fractions, with correlation coefficients of –0.72 ($p < 0.01$) and –0.54 ($p < 0.01$), respectively.

To enhance the prediction accuracy of the machine learning (ML) models, the dataset excluding the N2 concentration, moisture content (MC), fixed carbon (FC), and oxygen content (O2) features (Case 2) was used. Although the MC feature showed no significant relationship with other features, it was excluded because biomass is typically dried before torrefaction, making its moisture content unlikely to influence the model's predictions. Additionally, to better understand the contribution of specific features to prediction accuracy, a dataset excluding carbon content (C), hydrogen content (H2), nitrogen content (N2), and oxygen content (O2) features (Case 3) was also used for training and testing the models.

By analyzing these datasets, the study aimed to identify the most influential features and optimize the ML models for predicting the yield of torrefied biomass. This

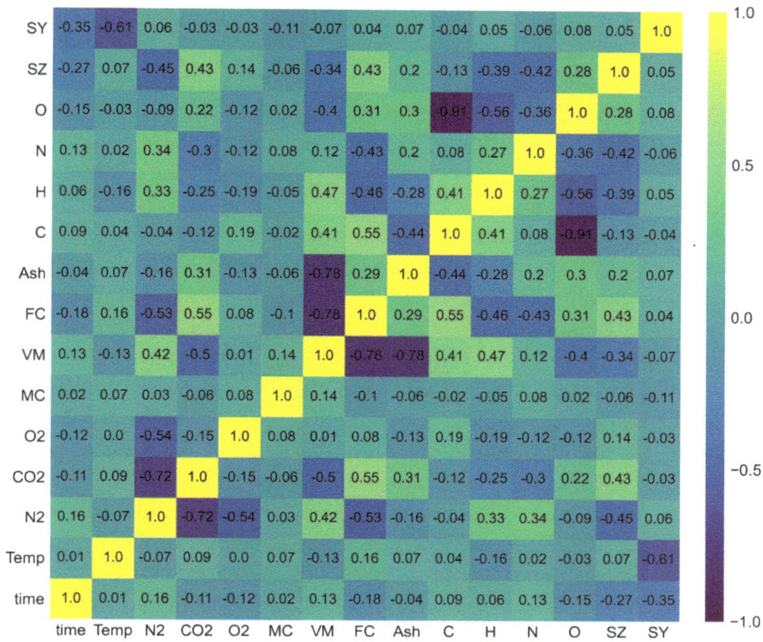

Fig. 4.1 Two variable Pearson correlation matrix from Onsree and Tippayawong (2021) (with permission)

approach not only improves the accuracy of the predictions but also provides valuable insights into the key factors driving the torrefaction process.

Using the optimal hyperparameters, the accuracy of the kernel ridge regression (KRR) and gradient tree boosting (GTB) models—measured by R^2, RMSE, and NRMSE—was evaluated through tenfold cross-validation for three different dataset cases, as shown in Table 4.2. Both models showed noticeable changes in accuracy

Table 4.2 Prediction accuracy in terms of R^2, RMSE, and NRMSE from tenfold cross-validation of both KRR and GTB algorithms for all three cases of input features, from Onsree and Tippayawong (2021) (with permission)

Model	Case	R^2	RMSE	NRMSE
KRR	Case 1	0.8618 (0.0398)	0.0771 (0.0085)	0.0990 (0.0114)
	Case 2	0.8731 (0.0177)	0.0778 (0.0090)	0.0996 (0.0152)
	Case 3	0.7778 (0.0203)	0.1050 (0.0059)	0.1526 (0.0110)
GTB	Case 1	0.8713 (0.0219)	0.0723 (0.0077)	0.0929 (0.0133)
	Case 2	0.8977 (0.0210)	0.0703 (0.0062)	0.0901 (0.0076)
	Case 3	0.8876 (0.0238)	0.0745 (0.0088)	0.1083 (0.0147)

Note These values were averaged from the results of tenfold cross-validation, in which the number in bracket was their standard deviation value. For all three datasets, the best hyperparameters of each model were optimized by a grid search with a nested tenfold cross-validation approach when comparing the full fourteen-feature dataset (Case 1) with datasets where certain features were removed (Cases 2 and 3).

For example, in Cases 1 and 2, KRR's R^2 increased slightly from 0.86 to 0.87, while GTB demonstrated a more significant improvement, with R^2 rising from approximately 0.87 to nearly 0.90. Both models maintained similar RMSE and NRMSE values across these cases, with KRR achieving values around 0.08 and 0.10, and GTB achieving values of approximately 0.07 and 0.09.

In Case 3, where the carbon content (C), hydrogen content (H2), nitrogen content (N2), and oxygen content (O2) features were removed, KRR's prediction accuracy declined significantly. Its R^2 dropped to about 0.78, while RMSE and NRMSE increased by over 50% compared to Case 1, reaching values of approximately 0.11 and 0.15, respectively. GTB also experienced a decline in performance, with R^2 falling to slightly above 0.88, and RMSE and NRMSE increasing to around 0.075 and 0.11, respectively.

An example for the distribution of prediction data for KRR and GTB models is shown in Fig. 4.2, in which the test data were from the first fold in ten folds of cross-validation for all three different sets of input features. It can be seen that the distribution of most data was inside a 10% error band for all cases. In summary, the GTB model, in particular, demonstrated a high level of accuracy in predicting the yield of solid products from biomass torrefaction, achieving an R^2 of about 0.90 and an average error of 0.070 w/w. This level of precision is comparable to models developed specifically for individual biomass types (Chiou et al. 2015) and aligns with similar applications of machine learning in other biomass-related processes (Garcia Nieto et al. 2019; Zhu et al. 2019). These results highlight the effectiveness of GTB in capturing the complex relationships between input features and torrefaction outcomes, making it a reliable tool for optimizing the torrefaction process.

These findings demonstrate that machine learning (ML) models can effectively predict the yield of torrefied biomass based on raw biomass properties and specific torrefaction conditions. Prediction accuracy can be further enhanced by including more relevant features in the dataset or removing features that are strongly correlated with each other. In this study, the gradient tree boosting (GTB) model achieved the highest accuracy for predicting torrefied biomass yields. The most influential features in the dataset were time, temperature, CO2 concentration, O2 concentration, volatile matter (VM), ash content, carbon content (C), hydrogen content (H2), nitrogen content (N2), and sample size (SZ) (Case 2).

To better understand how these features contribute to the GTB model's predictions, a ranking of feature importance was generated using the permutation importance method. As shown in Fig. 4.3, torrefaction temperature was identified as the most critical feature, with residence time being the second most important. Together, these two factors accounted for over 65% of the influence on solid yields (SYs). While this aligns with the basic Pearson correlation results, it confirms that the GTB model, combined with the permutation importance method, effectively identifies and highlights the key parameters driving the torrefaction process.

Beyond temperature and time, the next three most important features contributed approximately 20% to the solid yields. These features were related to the raw biomass

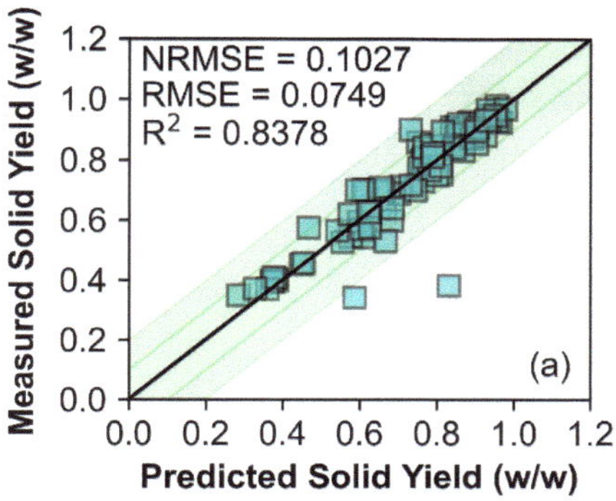

(a) Case 1: Cross validation for KRR

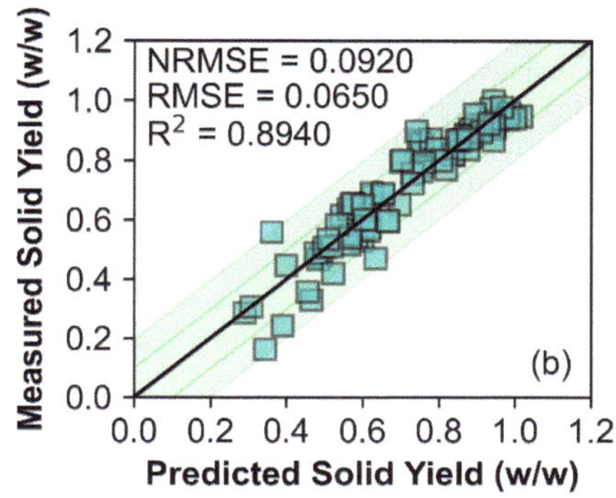

(b) Case 2: Cross validation for KRR

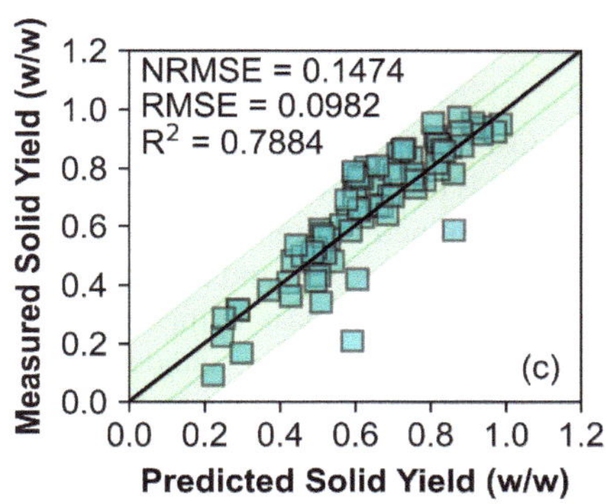

(c) Case 3: Cross validation for KRR

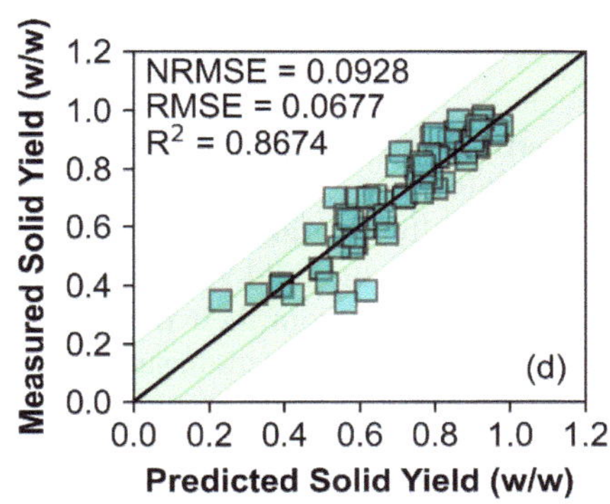

(d) Case 1: Cross validation for GTB

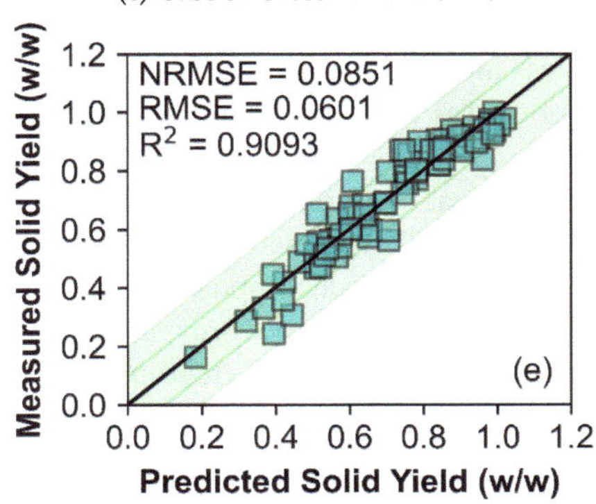

(e) Case 2: Cross validation for GTB

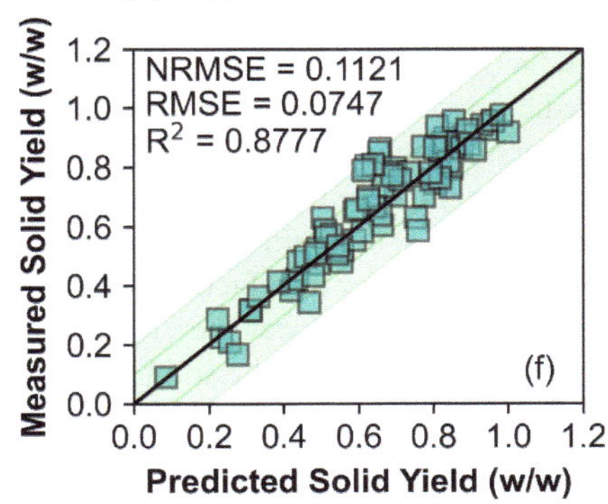

(f) Case 3: Cross validation for GTB

Fig. 4.2 Distribution of the prediction data using the test dataset from the first fold in tenfold of cross validation for KRR in (**a**), (**b**), and (**c**) as well as GTB in (**d**), (**e**), and (**f**) for Case 1 in (**a**) and (**d**), (ii) Case 2 in (**b**) and (**e**), and (iii) Case 3 in (**c**) and (**f**). The black trendline represents the positions of that the predicted values are exactly equal to the test values. The inner light green band indicates a 10% error range, and the outer dark green band indicates a 20% error range. (from Onsree and Tippayawong (2021), with permission)

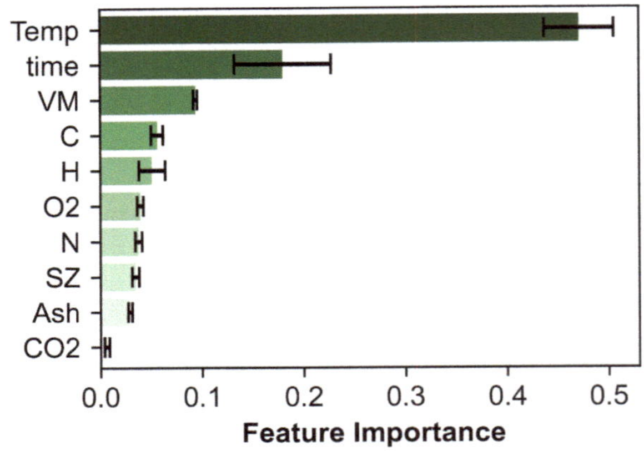

Fig. 4.3 Feature importance according to the GTB model from Onsree and Tippayawong (2021) (with permission)

properties, with volatile matter (VM) being the most significant, accounting for nearly half of this influence. Carbon (C) and hydrogen (H2) content followed, contributing almost equally. Overall, the biomass properties examined in this study contributed roughly 30% to the solid yields. The importance of VM, carbon, and hydrogen content likely stems from their direct connection to major biomass components and reactions during torrefaction, such as dehydration, de-volatilization, and decarboxylation (Onsree and Tippayawong 2020). This connection explains why these features have a more significant impact on torrefied biomass yields compared to other biomass properties.

By identifying and prioritizing these key features, the study provides valuable insights into the factors that most influence torrefaction outcomes. This information can be used to optimize the torrefaction process and improve the accuracy of yield predictions, making ML models like GTB powerful tools for advancing biomass conversion technologies.

In general, hemicellulose, which is a highly reactive component of biomass, decomposes at temperatures between 220 °C and 315 °C (Tippayawong et al. 2021). As a result, higher torrefaction temperatures lead to greater decomposition of hemicellulose, which reduces the yield of solid yields (SYs). In other words, as the temperature (T) increases, the solid yield tends to decrease. Similarly, extending the residence time of the torrefaction process also reduces the solid yield, though this effect is most pronounced for residence times between 10 and 40 min. Biomass requires a certain amount of time for heat to penetrate and break down unstable compounds (Onsree et al. 2018b). However, if the residence time exceeds 40 min, it has little to no additional effect on the yield.

The thermal decomposition of cellulose and lignin, the main components of biomass, occurs at temperatures above 300 °C (Barskov et al. 2019). Carbon content in raw biomass is often linked to the levels of cellulose and lignin (Yue et al. 2017). Therefore, biomass with higher cellulose and lignin content typically has lower hemicellulose content, which results in higher yields of torrefied biomass.

The effect of oxygen (O_2) in the reacting gas on solid yields is more complex. As the concentration of O_2 increases, solid yields tend to decrease. This is because O_2 promotes the partial oxidation of carbon in the biomass, converting it into carbon monoxide (CO). This process reduces both the carbon content in the torrefied biomass and the final yield. Understanding these relationships helps clarify how different factors influence the torrefaction process and provides insights for optimizing conditions to achieve desired outcomes.

Partial Dependence Plots (PDPs) were used to visualize the general impacts of key features on the target output after identifying their importance. Figure 4.4 presents one-way PDP results for the six most influential variables. The tick marks on the x-axis indicate the density of feature data, which helps give meaning to the trendline. Without these tick marks, interpreting the trends would be challenging.

In Fig. 4.4a, the T feature's influence on SYs shows a linear decrease as torrefaction temperature increases. Beyond 240 °C, this trend becomes slightly steeper. Hemicellulose, a highly reactive component of biomass, decomposes thermally between 220 °C and 315 °C (Tippayawong et al. 2021). This means higher torrefaction temperatures lead to greater decomposition of hemicellulose, resulting in lower SYs. In simpler terms, higher values of the T feature significantly reduce SYs.

Similarly, extending torrefaction residence time decreases the yield of solid products. As shown in Fig. 4.4b, this feature has a strong linear relationship with SYs for residence times between 10 and 40 min. Biomass requires a finite amount of time to transfer heat internally and decompose unstable compounds (Onsree et al. 2018b). Beyond 40 min, however, additional residence time does not affect SYs.

Figure 4.4c and e illustrate the impact of VM and H features on SYs, both showing a nearly monotonically decreasing trend. During torrefaction, hemicellulose is primarily decomposed, releasing volatile compounds. Consequently, raw biomass with higher hemicellulose content produces lower yields of torrefied biomass. This suggests that both VM and H features are indirectly linked to hemicellulose content.

Figure 4.4d reveals that the C feature has a positive effect on SYs, with a trendline that consistently rises as the carbon content in raw biomass increases. Similar to the VM and H features, the C feature is likely linked to the amounts of cellulose and lignin present, both of which are more thermally resistant compared to hemicellulose. The thermal decomposition of cellulose and lignin primarily occurs at temperatures above 300 °C (Barskov et al. 2019). Additionally, higher carbon content in raw biomass is associated with increased levels of cellulose and lignin (Yue et al. 2017). Therefore, when there are greater amounts of cellulose and/or lignin, the relative content of hemicellulose is lower, which results in higher yields of torrefied biomass.

In Fig. 4.4f, we see that the influence of oxygen (O_2) in the reacting gas on SYs is not straightforward. As the concentration of O_2 increases in the torrefaction environment, SYs gradually decrease. This reduction occurs because O_2 reacts with carbon in the raw biomass, leading to partial oxidation that produces carbon monoxide (CO). As a result, this process reduces the carbon content in the torrefied biomass and ultimately lowers SYs.

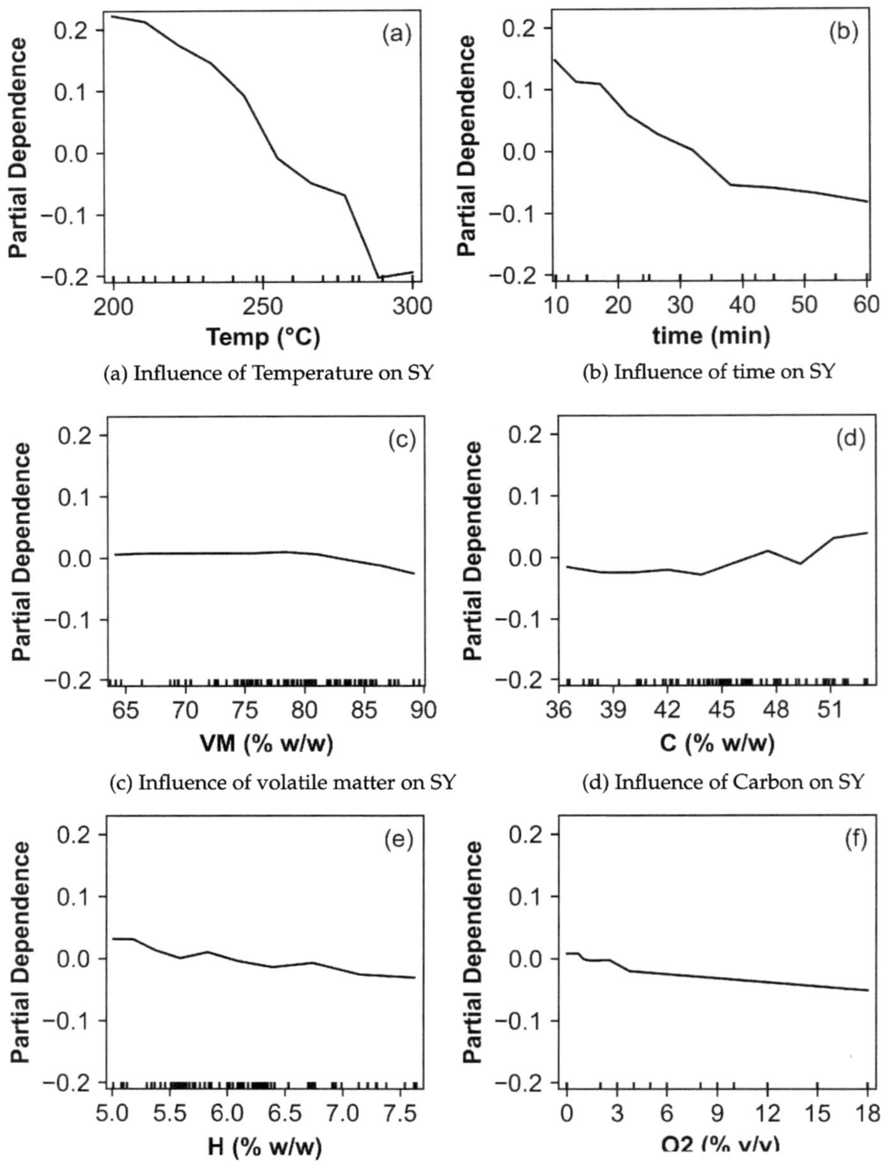

(a) Influence of Temperature on SY
(b) Influence of time on SY
(c) Influence of volatile matter on SY
(d) Influence of Carbon on SY
(e) Influence of Hydrogen on SY
(f) Influence of Oxygen on SY

Fig. 4.4 Partial dependence plots of the first six most important features by GTB model (from Onsree and Tippayawong (2021), with permission)

4.4 Conclusion

Machine learning techniques were successfully applied to predict the yield of solid products from biomass torrefaction processes, using both biomass properties and torrefaction conditions as input features. Among the tested algorithms, the gradient tree boosting (GTB) method demonstrated the highest prediction accuracy, achieving an R^2 value of nearly 0.90 and a root mean squared error (RMSE) of approximately 0.07.

In addition to torrefaction temperature and residence time, certain biomass properties—such as volatile matter, carbon content, and hydrogen content—played a significant role in predicting the yield, collectively contributing about 30% to the model's predictions. Volatile matter and hydrogen content were found to have a negative effect on the yield, while carbon content had a positive impact. These findings provide valuable insights into the complex relationships governing the biomass torrefaction process and highlight the importance of considering both process conditions and biomass characteristics when optimizing torrefaction outcomes.

By leveraging machine learning, this study offers a robust framework for predicting torrefied biomass yields and underscores the potential of data-driven approaches to enhance the efficiency and sustainability of biomass conversion technologies.

References

Alvarez A, Nogueiro D, Pizarro C, Matos M, Bueno JL (2018) Non-oxidative torrefaction of biomass to enhance its fuel properties. Energy 158:1–8. ISSN 0360-5442. https://doi.org/10.1016/j.energy.2018.06.009. https://www.sciencedirect.com/science/article/pii/S0360544218310624

Barskov S, Zappi M, Buchireddy P, Dufreche S, Guillory J, Gang D, Hernandez R, Bajpai R, Baudier J, Cooper R, Sharp R (2019) Torrefaction of biomass: a review of production methods for biocoal from cultured and waste lignocellulosic feedstocks. Renew Energy 142:624–642. ISSN 0960-1481. https://doi.org/10.1016/j.renene.2019.04.068. https://www.sciencedirect.com/science/article/pii/S096014811930552X

Bukhtiyarova M, Lunkenbein T, Kahler K, Schlogl R (2017) Methanol synthesis from industrial CO2 sources: a contribution to chemical energy conversion. Catal Lett 147(2):416–427. ISSN 1572-879X. https://doi.org/10.1007/s10562-016-1960-x

Cahyanti M, Doddapaneni T, Kikas T (2020) Biomass torrefaction: an overview on process parameters, economic and environmental aspects and recent advancements. Bioresour Technol 301:122737. ISSN 0960-8524. https://doi.org/10.1016/j.biortech.2020.122737. https://www.sciencedirect.com/science/article/pii/S0960852420300067

Cardona S, Gallego LJ, Valencia V, Martinez E, Rios LA (2019) Torrefaction of eucalyptus-tree residues: a new method for energy and mass balances of the process with the best torrefaction conditions. Sustain Energy Technol Assess 31:17–24. ISSN 2213-1388. https://doi.org/10.1016/j.seta.2018.11.002. https://www.sciencedirect.com/science/article/pii/S2213138818300882

Chiou B-S, Valenzuela-Medina D, Bilbao-Sainz C, Klamczynski AK, Avena-Bustillos RJ, Milczarek RR, Du W-X, Glenn GM, Orts WJ (2015) Torrefaction of pomaces and nut shells. Bioresour Technol 177:58–65. ISSN 0960-8524. https://doi.org/10.1016/j.biortech.2014.11.071. https://www.sciencedirect.com/science/article/pii/S0960852414016733

Chumsawat L, Tippayawong N (2020) Utilizing tobacco residues to generate bio-oil and biochar via ablative pyrolysis. Chem Eng Trans 78:49–54

Elmaz F, Yucel O, Mutlu AY (2020) Predictive modeling of biomass gasification with machine learning-based regression methods. Energy 191:116541. ISSN 0360-5442. https://doi.org/10.1016/j.energy.2019.116541. https://www.sciencedirect.com/science/article/pii/S0360544219322364

Fan Y, Tippayawong N, Wei G, Huang Z, Zhao K, Jiang L, Zheng A, Zhao Z, Li H (2020) Minimizing tar formation whilst enhancing syngas production by integrating biomass torrefaction pretreatment with chemical looping gasification. Appl Energy 260:114315. ISSN 0306-2619. https://doi.org/10.1016/j.apenergy.2019.114315. https://www.sciencedirect.com/science/article/pii/S0306261919320021

Fisher A, Rudin C, Dominici F (2019) All models are wrong, but many are useful: learning a variable's importance by studying an entire class of prediction models simultaneously. arXiv:1801.01489

Garcia Nieto PJ, Garcia-Gonzalo E, Paredes-Sanchez JP, Bernardo Sanchez A, Menendez Fernandez M (2019) Predictive modelling of the higher heating value in biomass torrefaction for the energy treatment process using machine-learning techniques. Neural Comput Appl 31(12):8823–8836. ISSN 1433-3058. https://doi.org/10.1007/s00521-018-3870-x

Kanwal S, Chaudhry N, Munir S, Sana H (2019) Effect of torrefaction conditions on the physicochemical characterization of agricultural waste (sugarcane bagasse). Waste Manag 88:280–290. ISSN 0956-053X. https://doi.org/10.1016/j.wasman.2019.03.053. https://www.sciencedirect.com/science/article/pii/S0956053X19301849

Khuenkaeo N, MacQueen B, Onsree T, Daiya S, Tippayawong N, Lauterbach J (2020) Bio-oils from vacuum ablative pyrolysis of torrefied tobacco residues. RSC Adv 10:34986–34995. https://doi.org/10.1039/D0RA06014C

Kuhn M, Johnson K (2013) Applied predictive modeling. ISBN 978-1-4614-6848-6. https://doi.org/10.1007/978-1-4614-6849-3

Liu Z, Han G (2015) Production of solid fuel biochar from waste biomass by low temperature pyrolysis. Fuel 158:159–165. ISSN 0016-2361. https://doi.org/10.1016/j.fuel.2015.05.032. https://www.sciencedirect.com/science/article/pii/S0016236115005281

Mohd Faizal H, Shamsuddin HS, Heiree MHM, Muhammad Ariff Hanaffi MF, Abdul Rahman MR, Rahman MM, Latiff Z (2018) Torrefaction of densified mesocarp fibre and palm kernel shell. Renew Energy 122:419–428. ISSN 0960-1481. https://doi.org/10.1016/j.renene.2018.01.118. https://www.sciencedirect.com/science/article/pii/S0960148118301289

Onsree T, Sittisun P, Sasaki R, Tippayawong N (2018a) Pyrolysis of corn residues: kinetic analysis using discrete distributed activation energy model. In: IOP conference series: earth and environmental science volume 159, Da Nang, Vietnam. 2018 4th international conference on environment and renewable energy (ICERE 2018)

Onsree T, Tippayawong N (2017) Application of gaussian smoothing technique in evaluation of biomass pyrolysis kinetics in macro-tga. Energy Procedia 138:778–783. ISSN 1876-6102. https://doi.org/10.1016/j.egypro.2017.10.059. https://www.sciencedirect.com/science/article/pii/S1876610217350002. 2017 international conference on alternative energy in developing countries and emerging economies

Onsree T, Tippayawong N (2020) Torrefaction of maize residue pellets with dry flue gas. BioEnergy Research 13(1):358–368. ISSN 1939-1242. https://doi.org/10.1007/s12155-019-10058-x

Onsree T, Tippayawong N (2021) Machine learning application to predict yields of solid products from biomass torrefaction. Renew Energy 167:425–432. ISSN 0960-1481. https://doi.org/10.1016/j.renene.2020.11.099. https://www.sciencedirect.com/science/article/pii/S0960148120318516

Onsree T, Tippayawong N, Zheng A, Li H (2018b) Pyrolysis behavior and kinetics of corn residue pellets and eucalyptus wood chips in a macro thermogravimetric analyzer. Case Stud Therm Eng 12:546–556. ISSN 2214-157X. https://doi.org/10.1016/j.csite.2018.07.011. https://www.sciencedirect.com/science/article/pii/S2214157X18301424

Pedregosa F, Varoquaux G, Gramfort A, Michel V, Thirion B, Grisel O, Blondel M, Prettenhofer P, Weiss R, Dubourg V et al (2011) Scikit-learn: Machine learning in python. J Mach Learn Res 12:2825–2830

Piboon P, Tippayawong N, Wongsiriamnuay T (2017) Densification of corncobs using algae as a binder. Chiang Mai Univ J Nat Sci 16. https://doi.org/10.12982/CMUJNA.2017.0014

Senliol B, Gulgezen G, Yu L, Cataltepe Z (2008) Fast correlation based filter (FCBF) with a different search strategy. In: 2008 23rd international symposium on computer and information sciences, pp 1–4. https://doi.org/10.1109/ISCIS.2008.4717949

Singh RK, Chakraborty JP, Sarkar A (2020) Optimizing the torrefaction of pigeon pea stalk (cajanus cajan) using response surface methodology (rsm) and characterization of solid, liquid and gaseous products. Renew Energy 155:677–690. ISSN 0960-1481. https://doi.org/10.1016/j.renene.2020.03.184. https://www.sciencedirect.com/science/article/pii/S0960148120305176

Tian X, Dai L, Wang Y, Zeng Z, Zhang S, Jiang L, Yang X, Yue L, Liu Y, Ruan R (2020) Influence of torrefaction pretreatment on corncobs: A study on fundamental characteristics, thermal behavior, and kinetic. Bioresour Technol 297:122490. ISSN 0960-8524. https://doi.org/10.1016/j.biortech.2019.122490. https://www.sciencedirect.com/science/article/pii/S0960852419317201

Tippayawong N, Jaipa C, Kwanseng K (2018) Biomass pellets from densification of tree leaf waste with algae. Agric Eng Int CIGR J 20(4):119–125

Tippayawong N, Onsree T, Williams T, McCullough K, MacQueen B, Lauterbach J (2021) Catalytic torrefaction of pelletized agro-residues with cu/al2o3 catalysts. Biomass Convers Biorefinery 11(5):1847–1852. ISSN 2190-6823. https://doi.org/10.1007/s13399-019-00535-w

Williams T, McCullough K, Lauterbach JA (2020) Enabling catalyst discovery through machine learning and high-throughput experimentation. Chem Mater 32(1):157–165. ISSN 0897-4756. https://doi.org/10.1021/acs.chemmater.9b03043

Wongsiriamnuay T, Tippayawong N (2015) Effect of densification parameters on the properties of maize residue pellets. Biosyst Eng 139:111–120. ISSN 1537-5110. https://doi.org/10.1016/j.biosystemseng.2015.08.009. https://www.sciencedirect.com/science/article/pii/S1537511015001506

Yeo I-K, Johnson R (2000) A new family of power transformations to improve normality or symmetry. Biometrika 87:12. https://doi.org/10.1093/biomet/87.4.954

Yue Y, Singh H, Singh B, Mani S (May2017) Torrefaction of sorghum biomass to improve fuel properties. Biores Technol 232:372–379

Zhang Z, Duan H, Zhang Y, Guo X, Yu X, Zhang X, Rahman MM, Cai J (2020) Investigation of kinetic compensation effect in lignocellulosic biomass torrefaction: Kinetic and thermodynamic analyses. Energy 207:118290. ISSN 0360-5442. https://doi.org/10.1016/j.energy.2020.118290. https://www.sciencedirect.com/science/article/pii/S0360544220313979

Zhu X, Li Y, Wang X (2019) Machine learning prediction of biochar yield and carbon contents in biochar based on biomass characteristics and pyrolysis conditions. Bioresour Technol 288:121527. ISSN 0960-8524. https://doi.org/10.1016/j.biortech.2019.121527. https://www.sciencedirect.com/science/article/pii/S0960852419307576

Chapter 5
Co-pyrolysis—Biomass and Plastic Waste

Abstract This chapter explores the process of co-pyrolyzing biomass with plastic waste to produce bio-oil. The amount of bio-oil generated, known as bio-oil yield (BOY), can vary significantly when plastics are added to biomass due to interactions between the two materials. However, understanding these interactions and accurately predicting BOY while considering multiple factors is complex and often limited by traditional methods. To tackle this challenge, this chapter employs a machine learning model called XGBoost, combined with a tool known as Shapley additive explanation (SHAP), to create clear and interpretable models for predicting both BOY and interaction effects during co-pyrolysis. SHAP is a method for explaining the output of any machine learning model by attributing each feature's contribution to the final prediction, using a game-theoretic approach based on Shapley values. The models are based on 26 input features. To address imbalances in the dataset, where certain outcomes were underrepresented, a technique called synthetic minority over-sampling was applied. The XGBoost models achieved impressive accuracy, with nearly 90% for BOY predictions and over 85% for predicting interaction effects. Using SHAP, the chapter provides insights into how individual features and their interactions influence the model's predictions. While reaction temperature and the biomass-to-plastic ratio were the most influential factors, feedstock characteristics overall contributed more than 60% to the co-pyrolysis process. These findings not only shed light on the mechanisms of co-pyrolysis but also highlight potential areas for future research and optimization. By combining machine learning with interpretability tools, this chapter offers a powerful approach to understanding and improving the co-pyrolysis process.

Keywords AI · Bioenergy · Mixed biomass · Waste-to-energy · Regression

5.1 Introduction

The Need for Renewable Energy Industrial growth and increasing urbanization have caused a large surge in the world's need for energy. In fact, it is expected that energy demands will rise by 80% before 2050 (Wang et al. 2021a). Currently, we

N. Tippayawong et al., *Machine Learning Applications for Bioenergy Conversion*, Green Energy and Technology, https://doi.org/10.1007/978-981-96-8086-3_5

get most of our energy from petroleum-based fuels. However, using these fuels is damaging to the environment. Pollution, greenhouse gas emissions, and potential health risks are some of the downsides. The amount of carbon dioxide (CO_2) in Earth's atmosphere has increased from 285 ppm before industrialization to approximately 425 ppm now. Even if we were to stop emissions immediately, the planet would still continue to warm. Because of this, many countries are planning to become carbon-neutral by 2050. The goal is to keep the global temperature from rising more than 1.5–2.0 °C by 2100 (Chen 2021). Therefore, finding alternative, renewable, and sustainable energy sources is very important.

Biomass as a Renewable Resource Biomass is a plentiful renewable resource that can provide a continuous and sustainable source of energy (Onsree et al. 2019; Khanal et al. 2020). Bioenergy from biomass is considered carbon-neutral, which means it doesn't add to environmental pollution (Yiin et al. 2021). Biomass can be turned into liquid fuels that can replace petroleum. This is done through processes such as pyrolysis (Zhu et al. 2022b) and hydrothermal liquefaction (Saengsuriwong et al. 2021). Pyrolysis involves breaking down organic material at high temperatures without oxygen. This process creates biofuels in solid, liquid, and gas forms. However, bio-oil, which is a product of biomass pyrolysis, has some problems. It contains a lot of water and oxygen, which makes it unstable, corrosive, thick, and low in heating value (Khuenkaeo et al. 2020). Improving the quality of bio-oil is important for it to be used on a large scale and sold commercially.

Co-pyrolysis: An Approach to Improve Bio-Oil An interesting way to improve bio-oil is to co-pyrolyze biomass with materials that have more hydrogen compared to carbon, such as plastics. Plastics are used widely in many industries but don't break down easily. This leads to environmental problems due to waste accumulation. Plastics contain a lot of carbon and hydrogen but little oxygen. This makes them useful for co-pyrolysis with biomass (Gin et al. 2021). Plastics like polystyrene (PS) have a higher hydrogen content compared to biomass. When co-pyrolyzed, the hydrogen from plastics reacts with oxygen-rich biomass compounds, reducing oxygen levels in the bio-oil. This "hydrogen donation" effect raises the H/C ratio of the final product, making it chemically closer to conventional fuels like gasoline. For example, Nardella et al. (2022), discovered that co-pyrolyzing wood and plastic (with 30 wt% plastic) increased the carbon and hydrogen content in bio-oil while decreasing the oxygen content. This is due to synergistic effects (SE). In this chapter, synergistic effects (SE) refer to the change in bio-oil yields that occurs when biomass and plastic are mixed and co-pyrolyzed, compared to the yields from the individual pyrolysis of biomass and plastic waste alone.

Co-pyrolysis minimizes undesirable byproducts. For example, polycyclic aromatic hydrocarbons (PAHs) and oxygenated compounds, which are common in plastics-only pyrolysis oils, are reduced when biomass is introduced. This occurs because biomass components help break down complex plastic-derived molecules into simpler, less toxic compounds. Using waste plastics like non-recyclable polystyrene not only improves bio-oil but also addresses environmental challenges. Plastics that are unsuitable for traditional recycling can be re-purposed in co-pyrolysis, reducing landfill waste while producing energy-dense fuels. This dual benefit supports sustainable resource use and aligns with circular economy principles. Other studies, such as the one by Suriapparao et al. (2022b), have found that SE can increase bio-oil yield and make the bio-oil more selective for aromatic hydrocarbons. SE can also help with secondary pyrolytic reactions, which improve bio-oil quality by changing its chemical composition.

Factors Influencing Co-pyrolysis The synergistic effects and the yield of pyrolysis products are important in co-pyrolysis. However, they are affected by many factors, such as the characteristics of the biomass and plastic, the types of pyrolyzers, catalysts, and the operating conditions. This makes it difficult to understand and predict the results. Theoretical models have been used to predict product yields, but these models are often complex and take a lot of time. They usually focus on main factors like temperature, reaction time, and the ratio of biomass to plastic, but not as much on how these factors interact. This limits their reliability and usefulness for optimizing the process.

As discussed in previous chapters, machine learning (ML) can handle complex, non-linear relationships among many input variables. ML models can leverage existing data to predict outcomes accurately and identify patterns within the data, making it useful for applications like pyrolysis and other bio-chemical conversion processes (Khan et al. 2023). For instance, Ullah et al. (2021) applied ML to predict bio-oil yields (BOY) from biomass pyrolysis using 11 input features, achieving high prediction accuracy. Likewise, Li et al. (2022) used ML to predict biochar properties with over 90% accuracy, and Tang et al. (2021) found that pyrolysis conditions were significant predictors of gas yield.

In co-pyrolysis of biomass and plastic waste, ML can predict not only product yields but also SE. A study by Alabdrabalnabi et al. (2022) used XGBoost, an effective ML algorithm, to predict BOY and biochar yield from biomass and plastic pyrolysis with high accuracy. However, their model provided a limited view of feature importance and did not fully capture SE.

Machine Learning Models for Predicting Co-Pyrolysis Outcomes Recent advances in machine learning (ML) have enabled accurate predictions of bio-oil yield (BOY) and synergistic effects (SE) in biomass-plastic co-pyrolysis systems. By leveraging 26 input features—including biomass composition, plastic type, pyrolysis conditions, and material ratios—ML models like XGBoost can uncover complex, non-linear relationships that govern product yields and SE. These models address key limitations of traditional methods, such as oversimplified assumptions about reaction stages or incomplete feature interactions.

Dataset Preparation and Balancing Imbalanced datasets, common in pyrolysis studies due to experimental constraints, were addressed using the Synthetic Minority Over-Sampling Technique with Gaussian Noise (SMOGN). This approach generates synthetic data points for underrepresented cases, improving model generalization. For instance, rare combinations of high plastic ratios ($>50\%$) or low-temperature co-pyrolysis scenarios were augmented, ensuring robust predictions across diverse operating conditions.

Model Development and Validation The XGBoost algorithm was selected for its ability to handle high-dimensional data and provide feature importance rankings. Models were trained on a dataset of 1,200 experimental co-pyrolysis cases, validated through tenfold cross-validation. Key outcomes include:

- Bio-oil yield prediction: Achieved an R^2 of 0.93, outperforming earlier Random Forest models (R^2 ~0.85).
- Synergistic effect quantification: Predicted SE intensity with 89% accuracy by analyzing hydrogen transfer efficiency and oxygen reduction patterns.

Interpret-ability Through SHAP Analysis Shapley Additive Explanations (SHAP) revealed critical drivers of BOY and SE:

1. Plastic-to-biomass ratio: Contributions peak at 30–50% plastic, aligning with optimal SE observed in experimental studies. Higher ratios (>60%) reduce model confidence due to competing decomposition kinetics.
2. Feedstock properties: Biomass lignin content and plastic hydrogen-to-carbon ratio jointly account for 41% of BOY variance. For example, polystyrene's high hydrogen content enhances deoxygenation by 18% compared to polyethylene.
3. Process conditions: Final pyrolysis temperature (550–650 °C) and heating rate (10–20 °C/min) dominate gas-phase reactions, increasing aromatic hydrocarbons by up to 27%.

Advancing Process Optimization The models identified previously overlooked interactions, such as the role of biomass ash content in catalyzing plastic-derived hydrocarbon cracking. At 5–10 wt% ash, bio-oil energy density improves by 12–15 MJ/kg due to reduced oxygenated compounds. These insights enable targeted adjustments, such as blending high-ash agricultural residues with polystyrene to maximize SE.

Limitations and Future Directions While the XGBoost-SHAP framework provides actionable insights, challenges remain in extrapolating predictions to novel feedstock combinations (e.g., marine biomass with biodegradable plastics). Future integration with mechanistic models could enhance extrapolation accuracy while retaining interpretability.

This ML-driven approach demonstrates how data science can accelerate the development of sustainable co-pyrolysis systems, reducing reliance on trial-and-error experimentation. Therefore, this chapter develops interpretable ML models to predict both BOY and SE for co-pyrolysis of biomass and plastic waste, using 26 input features based on material characteristics and operating conditions. The accuracy of these XGBoost models was evaluated using tenfold cross-validation, and impactful feature patterns were visualized with SHAP to improve understanding of the high-dimensional relationships within the system.

5.2 Methodology

The dataset for this study on biomass-plastic co-pyrolysis was carefully constructed to ensure reliability and relevance for machine learning applications. Here's how the data preparation process unfolded.

Data Collection and Filtering Data came from 360 experimental cases across peer-reviewed studies in major journals like Elsevier, the American Chemical Society, and the Royal Society of Chemistry. Only studies using lignocellulosic biomass (e.g., rice husk, sugarcane bagasse) and common plastics (polyethylene, PVC, polystyrene) were included. During data curation, features that lacked sufficient data, had missing values, or showed no correlation with target variables (as determined by Pearson's correlation analysis in Sect. 2.7.1) were removed. This resulted in a curated dataset of 360 data points, representing 32 biomass types.

Key Input Features The final dataset included 26 standardized variables split into two categories:

1. Experimental Conditions

- Temperature (300–800 °C): Higher temperatures (>500 °C) promote secondary reactions that reduce oxygenated compounds in bio-oil.
- Biomass-to-plastic ratio (10–90 wt%): Ratios near 1:1 maximize synergistic hydrogen transfer.
- Heating rate (5–50 °C/min): Slower rates (<20 °C/min) improve bio-oil stability by extending volatile residence time.

2. Material Characteristics

- Proximate analysis: Moisture content (MC, wt%), volatile matter content (VM, wt%), fixed carbon content (FC, wt%), and ash content (Ash, wt%).
- Compositional analysis: Carbon (C, wt%), Hydrogen (H, wt%), Nitrogen (N, wt%), Oxygen (O, wt%), Sulfur (S, wt%), and Chlorine (Cl_P, wt%). It's worth noting that the chlorine (Cl_P) feature was retained despite its limited distribution and low correlation with target variables, as it is a primary component of polyvinyl chloride.

Data Standardization To standardize the 26 input features, Yeo-Johnson's non-linear transformation was applied to achieve a normal (Gaussian) distribution, as described in Sect. 2.7. Bio-oil definitions vary across the literature, but for this chapter, bio-oil is defined as the total liquid fraction captured during co-pyrolysis, with (y_{BOY}) calculated accordingly.

$$y_{BOY} = \frac{m_{liquid}}{M} x100 \tag{5.1}$$

where m_{liquid} is the mass of the total liquid fraction, and M is the initial mass of the sample on a dry basis. Synergistic effects (SE) is defined here as the change in bio-oil yields that occurs when biomass and plastic are mixed and co-pyrolyzed, compared to the yields from the individual pyrolysis of biomass and plastic waste alone. The synergistic effect (SE), denoted as y_{SE}, was calculated using the following formula:

$$y_{SE} = \frac{\left(y_{BOY,co-pyrolysis} - y_{BOY,individual}\right)}{y_{BOY,individual}} x100 \tag{5.2}$$

This calculation captures the additional yield generated through co-pyrolysis, indicating the extent of interaction between biomass and plastic that enhances or suppresses bio-oil production.

$$y_{BOY,individual} = \omega y_{BOY,B} + (1 - \omega) y_{BOY,P} \tag{5.3}$$

where $y_{BOY,co-pyrolysis}$ is the bio-oil yield (BOY) from the co-pyrolysis experiment of a biomass and plastic mixture, $y_{BOY,individual}$ is the BOY value from an individual pyrolysis experiment of the biomass ($y_{BOY,B}$) and plastic ($y_{BOY,P}$) waste, and ω is the biomass-to-total mass ratio (in wt%) applied in the co-pyrolysis experiment. As a result, when the biomass-to-total mass ratio (B:T) was either 0 or 100 wt%, the synergistic effect (SE) value was 0. These conditions accounted for about 40% of the entire dataset, indicating a non-normal distribution of SE values. This imbalance creates challenges for machine learning models, as most algorithms struggle to detect meaningful patterns when critical data points are underrepresented—a problem known as imbalanced regression.

To address this, researchers used the SMOGN technique (Synthetic Minority Over-sampling for Regression with Gaussian Noise, (Branco et al. 2017)). This method balances datasets by:

1. Identifying rare, high-impact cases: A "rarity threshold" (set here to 0.9) separates less relevant data from critical underrepresented cases.
2. Reducing common data points: Random under-sampling decreases over represented cases.
3. Generating synthetic data:

- SMOTER (Synthetic Minority Over-sampling Technique for Regression): Creates new examples by blending features of similar nearby data points (like mixing colors on a palette).
- Gaussian Noise: Adds controlled randomness to distant data pairs, mimicking natural experimental variations.

For this work, the Python SMOGN library (from https://pypi.org/project/smogn/) was configured with:

- 8 neighboring data points for similarity comparisons
- 20% noise level for diversity

This generated 60 new bio-oil yield (BOY) and 140 SE data points, improving model training without distorting original trends. The approach mirrors how chefs adjust recipes—reducing abundant ingredients (under-sampling) while carefully enhancing rare flavors (targeted over-sampling) to create a balanced dish. By addressing data imbalances, SMOGN helps models better predict both standard and exceptional experimental outcomes in co-pyrolysis research.

Figure 5.1 demonstrates how using the SMOGN technique improves the ability of XGBoost (XGB) models to predict bio-oil yield (BOY) and synergistic effects (SE).

Bio-Oil Yield (BOY) Prediction Applying SMOGN increased the R^2 value for the test dataset from approximately 0.82 to around 0.88, as shown in panels (a) and (d) of Fig. 5.1. This means the model's predictions became more accurate overall. Panels (b) and (e) illustrate that SMOGN made the data distribution more balanced, bringing the training and test datasets into closer alignment. The R^2 value for the training data decreased with SMOGN, suggesting less over fitting and better generalization by the model. Panels (c) and (f) show that SMOGN resulted in a broader distribution of relative errors in the training dataset and a sharper distribution in the test dataset, further reducing over fitting.

Synergistic Effects (SE) Prediction SMOGN improved the accuracy for the test dataset from approximately $R^2 = 0.70$ to $R^2 = 0.84$, as seen in panels (g) and (j). Although the R^2 values for both training datasets remained high (above 0.99), indicating some over fitting, the additional data generated by SMOGN enhanced model performance. The increased data in the ranges of –40 to –10 and 10–40 (panels (h) and (k)) enabled the XGB model to better identify and learn patterns in the SE data, thereby improving prediction accuracy.

5.3 Model Performance Evaluation and Feature Selection

To make sure the machine learning (ML) models were accurate and unbiased, a careful evaluation process was used:

1. Tenfold Cross-Validation:

- The entire dataset was randomly mixed and divided into 10 equal parts.
- The model was trained on 9 parts and tested on the remaining part.
- This process was repeated 10 times, each time using a different part as the test set.
- This method provides a reliable and unbiased estimate of how well the model performs on new data (similar to giving a student multiple practice tests before a final exam).

2. Hyperparameter Optimization:

- The model has several settings (hyperparameters) that need to be tuned for best performance.
- A full-factor grid search systematically tested all possible combinations of hyperparameter values, where all possible combinations of hyperparameters were searched over, while other hyperparameters of the models were kept at default values (Chen and Guestrin 2016).

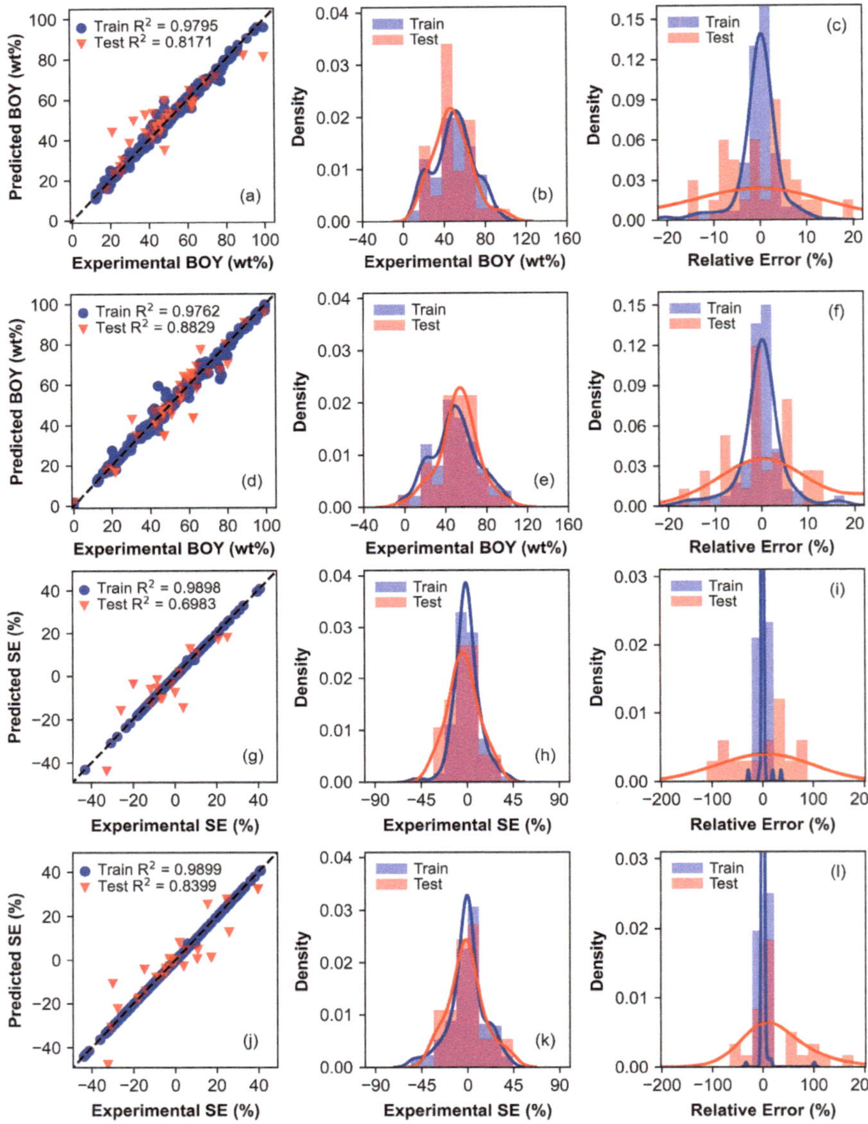

Fig. 5.1 Accuracy of the XGB model on making predictions of (a)-(f) BOY, without SMOGN (**a**)–(**c**) and (**g**)–(**l**) SE, without SMOGN (**g**)–(**i**), from Prasertpong et al. (2023) (with permission)

- Nested tenfold cross-validation was used to evaluate each hyperparameter set, selecting the combination that gave the best performance in predicting bio-oil yield (BOY) and synergistic effects (SE) (Katongtung et al. 2022; Wongchai et al. 2022).
- Model performance was measured using the coefficient of determination (R^2) and root mean squared error (RMSE) as defined in Eqs. 2.30 and 2.31.

3. Feature Importance Analysis: To understand which input features were most important for accurate predictions, the ML models were applied to different subsets of features:

- Case 0: All 26 features were used.
- Case 1: All features except co-pyrolysis experimental conditions.
- Case 2: All features except proximate analysis of biomass.
- Case 3: All features except proximate analysis of plastic waste.
- Case 4: All features except compositional analysis of biomass.
- Case 5: All features except compositional analysis of plastic waste.

By comparing the model performance with different feature subsets, the impact of each group of input features on the accuracy of predicting BOY and SE could be determined. This helped identify the most critical features for the models (like understanding which ingredients are essential in a recipe).

5.3.1 XGB Algorithm

eXtreme Gradient Boosting (XGB) was introduced in Sect. 2.6. In this chapter, a Python library of XGB from https://pypi.org/project/xgboost/ was applied, and the optimized hyperparameters of the XGB models for predictions of the BOY and SE are listed in Table 5.1.

5.4 Results and Discussion

Table 5.2 summarizes how different sets of input features affect the accuracy and learning time of XGBoost (XGB) models in predicting bio-oil yield (BOY) and synergistic effects (SE).

Case 0: All Features Using the complete 26-feature dataset resulted in the highest prediction accuracy for both BOY (approximately $R^2 \backsim 0.89$, RMSE = 6.81, and

Table 5.1 Optimized hyperparameters of the XGB models, from Prasertpong et al. (2023)(with permission)

Hyperparameters	Optimized values	
	BOY	SE
'Learning_rate'	0.15	0.30
'Subsample'	0.80	0.80
'Gamma'	0.15	0.000015
'Colsample_bytree'	0.20	0.20
'Max_depth'	6	6
'Max_leaves'	3	2

Table 5.2 Accuracy of making predictions of BOY and SE in Cases 0–5 from the XGB models, from Prasertpong et al. (2023) (with permission)

		BOY				SE			
		R^2	RMSE	NRMSE	Learning time (ms)	R^2	RMSE	NRMSE	Learning time (ms)
Case 0	Train	0.9891 (0.0021)	2.2850 (0.2001)	0.0462 (0.0041)	239.1 (26.2)	0.9998 (0.0001)	0.1179 (0.0575)	–0.1473 (0.0822)	251.6 (10.9)
	Test	0.8901 (0.0546)	6.8106 (1.3105)	0.1395 (0.0349)	237.5 (26.9)	0.8535 (0.0357)	6.5833 (0.9193)	2.3275 (28.6615)	246.9 (9.4)
Case 1	Train	0.5491 (0.0085)	14.7965 (0.1949)	0.2991 (0.0040)	225.1 (40.3)	0.7765 (0.0141)	8.3205 (0.2995)	–9.682 (2.6411)	232.8 (17.8)
	Test	0.5272 (0.0991)	17.4628 (1.7336)	0.354 (0.0353)	223.5 (35.7)	0.6209 (0.0855)	11.8204 (2.4827)	0.7608 (57.1957)	232.8 (17.8)
Case 2	Train	0.9856 (0.002)	2.6376 (0.1555)	0.0533 (0.0033)	220.3 (4.7)	0.9998 (0.0001)	0.1184 (0.0577)	–0.1479 (0.0826)	248.5 (10.9)
	Test	0.8791 (0.0638)	7.1293 (1.3423)	0.1463 (0.0374)	220.3 (4.7)	0.8061 (0.0939)	7.4501 (2.0298)	5.8808 (29.1416)	243.8 (7.7)
Case 3	Train	0.9842 (0.0031)	2.7574 (0.2353)	0.0558 (0.0049)	215.6 (6.3)	0.9998 (0.0001)	0.1183 (0.0576)	–0.1478 (0.0825)	253.1 (6.3)
	Test	0.8695 (0.0575)	7.5161 (1.3979)	0.1538 (0.0367)	214.1 (7.2)	0.8388 (0.0822)	6.8065 (1.9252)	–0.4294 (28.421)	253.1 (6.3)
Case 4	Train	0.9864 (0.0015)	2.5693 (0.119)	0.0519 (0.0025)	228.4 (10.5)	0.9998 (0.0001)	0.1184 (0.0577)	–0.148 (0.0827)	250.1 (7.0)
	Test	0.8813 (0.0586)	7.0908 (1.3401)	0.1453 (0.0361)	225.2 (7.9)	0.8054 (0.0712)	7.5426 (1.636)	4.8005 (34.9698)	248.5 (8.4)
Case 5	Train	0.9855 (0.002)	2.6499 (0.1646)	0.0536 (0.0035)	221.9 (6.3)	0.9998 (0.0001)	0.1183 (0.0577)	–0.1478 (0.0825)	260.6 (18.2)
	Test	0.8545 (0.0569)	7.9532 (1.1206)	0.1628 (0.033)	220.3 (4.7)	0.8518 (0.0843)	6.4955 (2.0973)	–0.3829 (26.7961)	259.0 (18.3)

Note These values were averaged from the results of tenfold cross-validation, in which the number in bracket was their SD value. For all six datasets, the best hyperparameters of each model were optimized by full-factor grid search with nested tenfold cross-validation. The NRMSE for test datasets of SE had a large fluctuation because the Y for some folds of the test datasets was nearly 0

NRMSE = 0.14) and SE (approximately $R^2 \backsim 0.85$, RMSE = 6.58, and NRMSE = 2.33). The Normalized Root Mean Squared Error (NRMSE) for SE can be positive or negative, depending on the $\overline{Y}_s$.

Case 1: Excluding Co-pyrolysis Conditions Removing features related to co-pyrolysis conditions led to the lowest prediction accuracy, with $R^2 = 0.55$ for BOY and $R^2 = 0.78$ even for the training dataset. This shows how important co-pyrolysis conditions are for model accuracy.

Cases 2–5: Excluding Raw Material Features Removing specific feature groups related to raw materials only slightly reduced accuracy. For BOY predictions, R^2 values ranged from 0.85 to 0.88, RMSE values exceeded 7.0, and NRMSE values were above 0.14. For SE predictions, R^2 values ranged from 0.80 to 0.85, RMSE values were above 6.5, and NRMSE values were between –0.38 and 5.90.

Feature Importance Among the raw material features, compositional analysis had the most significant impact on predicting BOY, while proximate analysis was more influential for predicting SE.

Conclusion Including features related to co-pyrolysis conditions, compositional analysis (for BOY), and proximate analysis (for SE) improves model accuracy. These findings are consistent with previous research on specific biomass types and other ML models developed for alternative conversion methods (Alabdrabalnabi et al. 2022; Li et al. 2022; Tang et al. 2021), indicating that XGB models effectively predict BOY and SE for co-pyrolysis of biomass and plastic waste.

5.4.1 SHAP Feature Importance

Figure 5.2 illustrates the influence of various input features on bio-oil yield (BOY) and synergistic effects (SE) during co-pyrolysis, considering both experimental conditions and the characteristics of biomass and plastic waste. SHAP values are used to quantify each feature's impact: larger fluctuations indicate greater significance, with red and blue colors representing high and low feature values, respectively.

Key Influential Features For BOY, the eight most influential features include temperature (T), the biomass-to-plastic ratio (B:P), fixed carbon in plastic (FC_P), volatile matter in plastic (VM_P), and hydrogen content in plastic (H_P). For SE, the top features

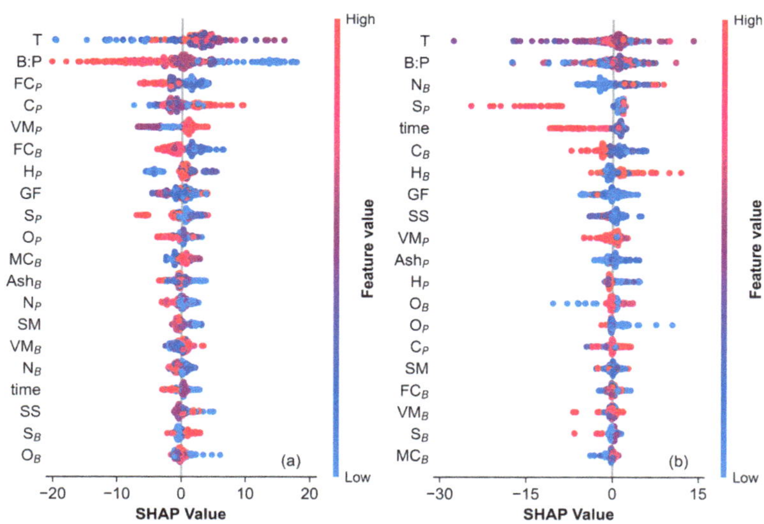

(a) Feature importance by SHAP value for BOY (b) Feature importance by SHAP value for SE

Fig. 5.2 Feature importance by SHAP value for (**a**) BOY and (**b**) SE, from Prasertpong et al. (2023) (with permission)

are T, B:P, nitrogen content in biomass (N_B), sulfur content in plastic (S_P), reaction time, and carbon balance (C_B).

Studies show that temperature and biomass-to-plastic ratio are key factors in co-pyrolysis outcomes (Hassan et al. 2016). Higher temperatures can enhance BOY, but excessive temperatures may lead to bio-oil cracking and reduced yields (Abnisa and Wan Daud 2014; Engamba Esso et al. 2022). Lower B:P ratios promote hydrogen transfer from plastic to bio-oil, increasing both BOY and SE, (Ahmed et al. 2020).

Unexpectedly Influential Features "Gas flow" (GF) unexpectedly influenced both BOY and SE. Although GF is part of the co-pyrolysis conditions, its impact is also related to reactor type (Alabdrabalnabi et al. 2022; Kumagai et al. 2016; Sophonrat et al. 2018). Higher GF values help reduce secondary bio-oil cracking reactions, potentially improving BOY and SE.

Feature Impacts on BOY Plastic-related features like carbon content in plastic (C_P), volatile matter in plastic (VM_P), and hydrogen content in plastic (H_P) are significant for BOY, as they are associated with hydrocarbons and hydrogen available during reactions. Conversely, features like fixed carbon in plastic (FC_P) and fixed carbon content in biomass (FC_B) contribute to char formation, increasing solid residues and decreasing BOY.

Feature Impacts on SE For SE, features such as hydrogen content in biomass (H_B), carbon content in biomass (C_B), nitrogen content in biomass (N_B), and sulfur content in plastic (S_P) are crucial. H_B influences hydrogen transfer between biomass and plastic, forming pyrolytic products (Sophonrat et al. 2018. C_B affects the release of volatiles from raw materials, with higher C_B limiting this release. The N_B feature is linked to nitrogen content, which can produce ammonia (NH_3) and hydrogen cyanide (HCN) during co-pyrolysis (Chen et al. 2020; Tang et al. 2018). The sulfur content (S_P) in plastic was identified as a key variable for the first time in this machine learning model.

Overall Contribution of Feature Groups Co-pyrolysis conditions account for 35% of the BOY prediction model and 30% for SE, based on SHAP values. This aligns with previous studies that identify heating rates as primary factors in co-pyrolysis of biomass and other materials (Brand et al. 2014; Alabdrabalnabi et al. 2022; Abnisa and Wan Daud 2014).

5.4.2 One- and Two-Way SHAP Feature Interpretation

Figures 5.3 and 5.4 use SHAP dependence plots to illustrate how the eight most influential features affect bio-oil yield (BOY) and synergistic effects (SE) based on the XGBoost (XGB) model. In these plots, each dot represents an instance, colored by another interacting feature to show nuances in the relationships. SHAP values indicate a feature's influence on the model's prediction relative to a baseline: positive values indicate a positive influence, negative values indicate a negative influence, and zero values indicate no change from the baseline.

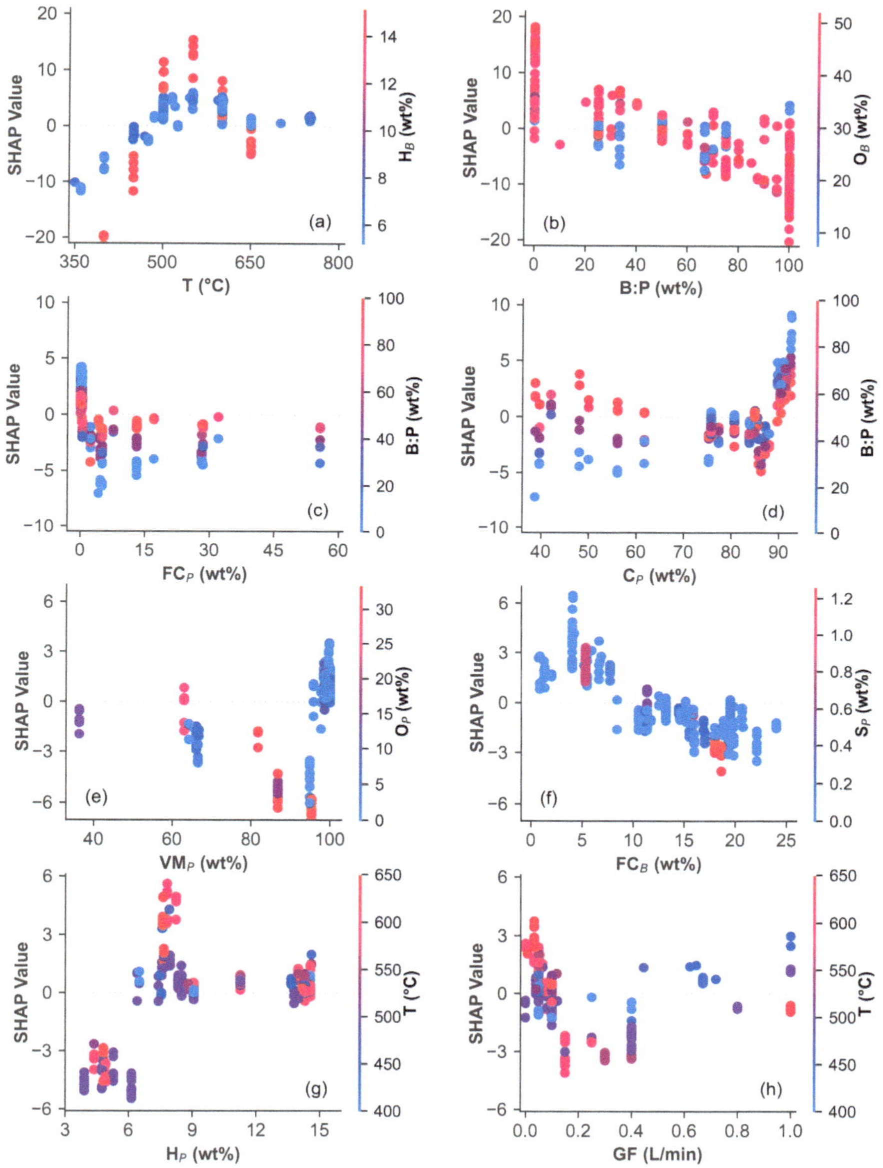

Fig. 5.3 SHAP dependence plots of the first eight most influential features on making predictions by the XGB model for BOY. A vertical color bar shows another feature that each of these eight features has the most interaction with: (**a**) the T feature with the H_B interaction, (**b**) the B:P feature with the O_B interaction, (**c**) the FC_P feature with the B:P interaction, (**d**) the C_P feature with the B:P interaction, (**e**) the VM_P feature with the O_P interaction, (**f**) the FC_B feature with the S_P interaction, (**g**) the H_P feature with the T interaction, and (**e**) the GF feature with the T interaction. Note: A dot represents each data point considered in this study, from Prasertpong et al. (2023) (with permission)

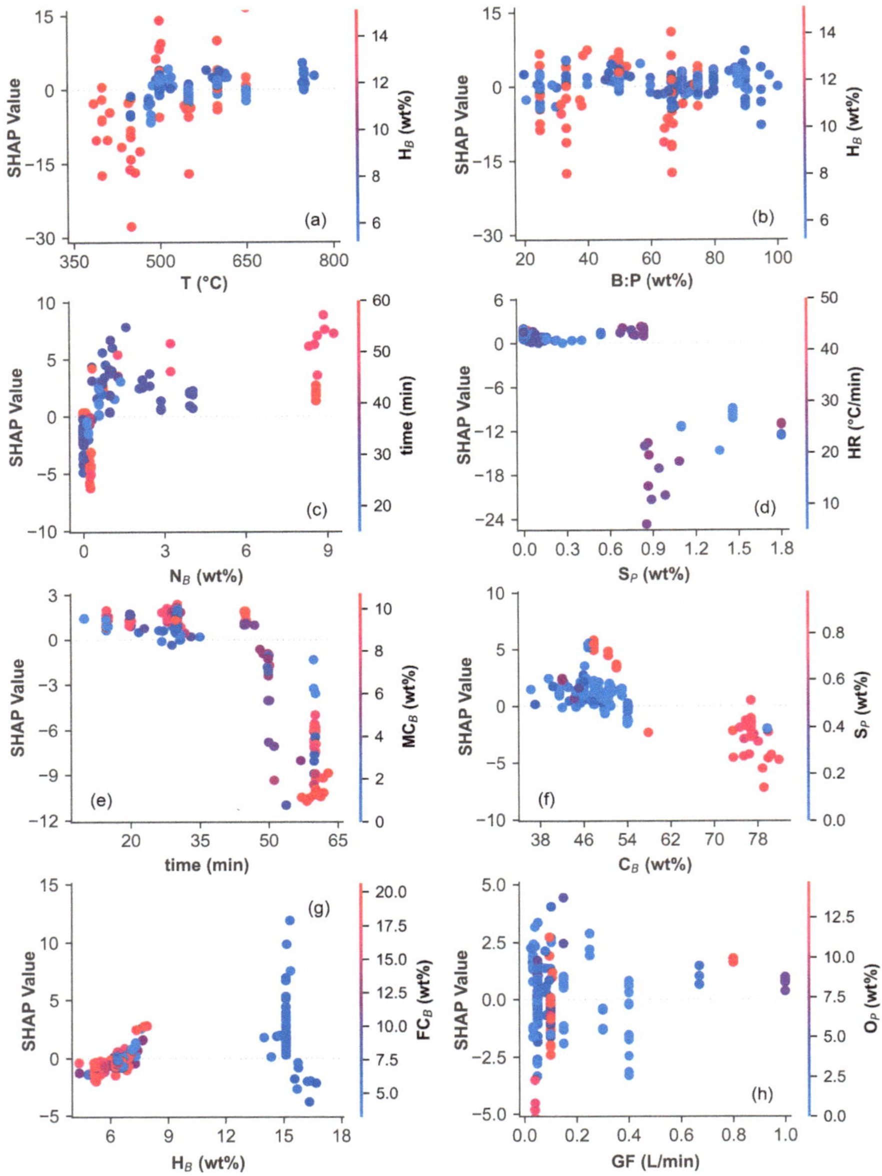

Fig. 5.4 SHAP dependence plots of the first eight most influent features on making predictions by the XGB model for SE. A vertical color bar shows another feature that each of these eight features has the most interaction with: (**a**) the T feature with the H_B interaction, (**b**) the B:P feature with the H_B interaction, (**c**) the N_B feature with the time interaction, (**d**) the S_P feature with the HR interaction, (**e**) the time feature with the MC_B interaction, (**f**) the C_B feature with the S_P interaction, (**g**) the H_B feature with the FC_B interaction, and (**e**) the GF feature with the O_P interaction. Note: A dot represents each data point considered in this study, from Prasertpong et al. (2023) (with permission)

Bio-Oil Yield (BOY) Features

(a) Temperature (T): In Fig. 5.3a, SHAP values generally increase with temperature up to approximately 550 °C, after which they decrease. This occurs because biomass and plastic undergo thermal decomposition between 250 °C and 550 °C, enhancing bio-oil production. However, at higher temperatures, cracking reactions become dominant, converting bio-oil to gaseous products. The T feature has a stronger positive impact between 480 °C and 620 °C when combined with high hydrogen balance (H_B) values, which promotes hydrogen transfer and increases bio-oil yield, (Wang et al. 2021a). This suggests that co-pyrolysis near 600 °C may be optimal.

(b) Biomass-to-Plastic Ratio (B:P): Fig. 5.3b shows that a low biomass-to-plastic ratio (B:P) tends to enhance bio-oil yield, particularly when the oxygen content in biomass (O_B) is high. Research indicates that pure plastic produces more bio-oil than pure biomass, (Sanahuja-Parejo et al. 2019). Conversely, a high B:P ratio can reduce the yield when the biomass has high oxygen content, as this decreases the hydrogen available for forming hydrocarbon liquids.

(c) Fixed Carbon in Plastic (FC_P): As shown in Fig. 5.3c, when the fixed carbon in plastic (FC_P) exceeds 5 wt%, SHAP values become negative, especially at low B:P values. This is because high fixed carbon tends to increase biochar formation, which reduces BOY during co-pyrolysis, (Abnisa and Wan Daud 2014; Ahmed et al. 2020).

(d) Carbon in Plastic (C_P): Fig. 5.3d demonstrates that SHAP values are strongly positive for the carbon in plastic (C_P) feature when it exceeds 90 wt%. This indicates that plastic with high volatile matter (VM) content enhances bio-oil production, (Ahmed et al. 2020).

(e) Volatile Matter in Plastic (VM_P): Similarly, high volatile matter in plastic (VM_P) also corresponds to positive SHAP values, as seen in Fig. 5.3e.

(f) Fixed Carbon in Biomass (FC_B): For fixed carbon in biomass (FC_B), Fig. 5.3f displays an inverse, linear relationship with SHAP values, similar to the trend observed for FC_P.

(g) Hydrogen content in plastic (H_P): Fig. 5.3g highlights that Hydrogen content in plastic (H_P) shows high SHAP values when the plastic's hydrogen content is between 7 and 9 wt% at elevated temperatures. This condition promotes hydrogen transfer, boosting BOY, (Alabdrabalnabi et al. 2022).

(h) Gas Flow (GF): Meanwhile, SHAP values for gas flow (GF) in Fig. 5.3h vary significantly, as GF aids vapor cooling, which can reduce vapor cracking and improve BOY depending on the reactor design.

Synergistic effects (SE)

(a) Figure 5.4a shows that higher temperatures (T) increase SHAP values, especially when the hydrogen balance (H_B) values are between 360 °C and 520 °C. Hydrogen transfer is more favorable in this temperature range, leading to the formation of hydrocarbons. However, SE decreases at very high temperatures because most volatiles have already been released.

(b) The SHAP values in Fig. 5.4b vary across the biomass-to-plastic ratio (B:P) feature. Similar to the T feature, the influence of B:P increases at high HB values. This suggests that hydrogen donation and the interaction between biomass and plastic volatile matter play a key role in forming hydrocarbon liquid products and enhancing SE, (Kumagai et al. 2016; Sophonrat et al. 2018).
(c) In Fig. 5.4c, high values of the nitrogen balance (NB) feature have a stronger positive effect on SE, particularly over longer reaction times.
(d) Conversely, in Fig. 5.4d, the SHAP value decreases as the sulfur presence (SP) feature increases, with a significant drop occurring once SP exceeds 0.7 wt%. This suggests that the optimum SP value for a positive SE effect should be no more than 0.7 wt%.
(e) Negative SHAP values are indicated with longer reaction times in Fig. 5.4e. In general, extended reaction times lead to secondary reactions where bio-oil cracks into non-condensable gases, reducing SE.
(f) Similarly, in Fig. 5.4f, the carbon balance (CB) feature has a negative impact on SE at high SP values, which aligns with earlier findings.
(g) Figure 5.4g shows higher SHAP values when HB values are high and fixed carbon in biomass (FCB) values are low. This indicates an interaction between hydrogen and the fixed carbon in biomass during co-pyrolysis with plastic waste. Using feedstocks with high hydrogen content can increase hydrocarbon liquid products, while low fixed carbon allows for greater volatile matter release, improving SE.
(h) The SHAP values fluctuate for the gas flow (GF) feature in Fig. 5.4h. The optimal gas flow rate depends on the type of pyrolysis reactor used. For instance, auger and fixed-bed reactors often need high GF. A higher gas flow rate can enhance heat and mass transfer between the feedstock and the pyrolysis environment, which can potentially improve SE.

5.5 Conclusion

This chapter has shown the individual and combined effects that different input features have on bio-oil yield (BOY) and synergistic effects (SE) when biomass and plastic waste are co-pyrolyzed. Although the models developed here were successful, a limited range of raw data meant that some important features, such as heating rate, were not fully captured. Future studies could expand the input data range to gain deeper insights. Catalytic co-pyrolysis, which can potentially increase bio-oil yield and improve its properties, is another exciting area for future research using machine learning approaches.

The interpretable XGB models developed in this chapter successfully predicted bio-oil yield (BOY) with an accuracy of $R^2 = 0.9$ and synergistic effects (SE) with $R^2 > 0.85$ for the co-pyrolysis of biomass and plastic waste. The SMOGN technique effectively addressed the imbalance in the training datasets.

Using SHAP analysis, it became clear that both individual features and interactions between features play key roles in the co-pyrolysis process. Temperature (T) and the biomass-to-plastic ratio (B:P) were among the most important features. Overall, the specific properties of the biomass and plastic waste influenced over 60% of the model's predictions. The combination of XGB and SHAP proved to be valuable for better understanding co-pyrolysis and shows great promise for supporting the future adoption of this technology.

References

Abnisa F, Wan Daud WMA (2014) A review on co-pyrolysis of biomass: an optional technique to obtain a high-grade pyrolysis oil. Energy Convers Manag 87:71–85. ISSN 0196-8904. https://doi.org/10.1016/j.enconman.2014.07.007. https://www.sciencedirect.com/science/article/pii/S019689041400630X

Ahmed MH, Batalha N, Mahmudul HM, Perkins G, Konarova M (2020) A review on advanced catalytic co-pyrolysis of biomass and hydrogen-rich feedstock: insights into synergistic effect, catalyst development and reaction mechanism. Bioresour Technol 310:123457. ISSN 0960-8524. https://doi.org/10.1016/j.biortech.2020.123457. https://www.sciencedirect.com/science/article/pii/S096085242030729X

Alabdrabalnabi A, Gautam R, Mani Sarathy S (2022) Machine learning to predict biochar and bio-oil yields from co-pyrolysis of biomass and plastics. Fuel 328:125303. ISSN 0016-2361. https://doi.org/10.1016/j.fuel.2022.125303. https://www.sciencedirect.com/science/article/pii/S001623612202138X

Andrade Cruz I, Chuenchart W, Long F, Surendra K, Renata Santos Andrade L, Bilal M, Liu H, Figueiredo RT, Khanal SK, Romanholo Ferreira LF (2022) Application of machine learning in anaerobic digestion: perspectives and challenges. Bioresour Technol 345:126433. ISSN 0960-8524. https://doi.org/10.1016/j.biortech.2021.126433. https://www.sciencedirect.com/science/article/pii/S0960852421017752

Branco P, Torgo L, Ribeiro RP (2017) Smogn: a pre-processing approach for imbalanced regression. In: First international workshop on learning with imbalanced domains: theory and applications, pp 36–50. PMLR

Brand S, Hardi F, Kim J, Suh DJ (2014) Effect of heating rate on biomass liquefaction: differences between subcritical water and supercritical ethanol. Energy 68:420–427. ISSN 0360-5442. https://doi.org/10.1016/j.energy.2014.02.086. https://www.sciencedirect.com/science/article/pii/S0360544214002205

Chen JM (2021) Carbon neutrality: toward a sustainable future. Innov 2(3):100127. ISSN 2666-6758. https://doi.org/10.1016/j.xinn.2021.100127. https://www.sciencedirect.com/science/article/pii/S2666675821000527

Chen T, Guestrin C (2016) Xgboost: a scalable tree boosting system. In: Proceedings of the 22nd ACM SIGKDD international conference on knowledge discovery and data mining, KDD 16, pp 785–794. Association for Computing Machinery, New York, NY, USA. ISBN 9781450342322. https://doi.org/10.1145/2939672.2939785

Chen R, Zhang S, Cong K, Li Q, Zhang Y (2020) Insight into synergistic effects of biomass-polypropylene co-pyrolysis using representative biomass constituents. Bioresour Technol 307:123243. ISSN 0960-8524. https://doi.org/10.1016/j.biortech.2020.123243. https://www.sciencedirect.com/science/article/pii/S0960852420305149

de Santi NM, Rodrigues NN, Montero-Dorta AD, Abramo LR, Tucci B, Artale MC (2022) Mimicking the halo galaxy connection using machine learning. Mon Not R Astron Soc 514(2):2463–2478. ISSN 1365-2966. https://doi.org/10.1093/mnras/stac1469

Engamba Esso SB, Xiong Z, Chaiwat W, Kamara MF, Longfei X, Xu J, Ebako J, Jiang L, Su S, Hu S, Wang Y, Xiang J (2022) Review on synergistic effects during co-pyrolysis of biomass and plastic waste: significance of operating conditions and interaction mechanism. Biomass Bioenergy

159:106415. ISSN 0961-9534. https://doi.org/10.1016/j.biombioe.2022.106415. https://www.sciencedirect.com/science/article/pii/S0961953422000769

Gin A, Hassan H, Ahmad M, Hameed B, Mohd Din A (2021) Recent progress on catalytic co-pyrolysis of plastic waste and lignocellulosic biomass to liquid fuel: the influence of technical and reaction kinetic parameters. Arab J Chem 14(4):103035. ISSN 1878-5352. https://doi.org/10.1016/j.arabjc.2021.103035. https://www.sciencedirect.com/science/article/pii/S1878535221000502

Hassan H, Lim J, Hameed B (2016) Recent progress on biomass co-pyrolysis conversion into high-quality bio-oil. Bioresour Technol 221:645–655. ISSN 0960-8524. https://doi.org/10.1016/j.biortech.2016.09.026. https://www.sciencedirect.com/science/article/pii/S0960852416312810

Huang N, Gao K, Yang W, Pang H, Yang G, Wu J, Zhang S, Chen C, Long L (2022) Assessing sediment organic pollution via machine learning models and resource performance. Bioresour Technol 361:127710. ISSN 0960-8524. https://doi.org/10.1016/j.biortech.2022.127710. https://www.sciencedirect.com/science/article/pii/S0960852422010392

Jankovic R, Mihajlovic I, Strbac N, Amelio A (2021) Machine learning models for ecological footprint prediction based on energy parameters. Neural Comput Appl 33(12):7073–7087. ISSN 1433-3058. https://doi.org/10.1007/s00521-020-05476-4

Kaczor Z, Bulinski Z, Werle S (2020) Modelling approaches to waste biomass pyrolysis: a review. Renew Energy 159:427–443. ISSN 0960-1481. https://doi.org/10.1016/j.renene.2020.05.110. https://www.sciencedirect.com/science/article/pii/S0960148120308089

Katongtung T, Onsree T, Tippayawong N (2022) Machine learning prediction of biocrude yields and higher heating values from hydrothermal liquefaction of wet biomass and wastes. Bioresour Technol 344:126278. ISSN 0960-8524. https://doi.org/10.1016/j.biortech.2021.126278. https://www.sciencedirect.com/science/article/pii/S0960852421016205

Khanal SK, Varjani S, Sze Ki Lin C, Awasthi MK (2020) Waste-to-resources: opportunities and challenges. Bioresour Technol 317:123987. ISSN 0960-8524. https://doi.org/10.1016/j.biortech.2020.123987. https://www.sciencedirect.com/science/article/pii/S0960852420312591

Khan M, Raza Naqvi S, Ullah Z, Ali Ammar Taqvi S, Nouman Aslam Khan M, Farooq W, Taqi Mehran M, Juchelkova D, Stepanec L (2023) Applications of machine learning in thermochemical conversion of biomass-a review. Fuel 332:126055. ISSN 0016-2361. https://doi.org/10.1016/j.fuel.2022.126055. https://www.sciencedirect.com/science/article/pii/S0016236122028794

Khuenkaeo N, MacQueen B, Onsree T, Daiya S, Tippayawong N, Lauterbach J (2020) Bio-oils from vacuum ablative pyrolysis of torrefied tobacco residues. RSC Adv 10:34986–34995. https://doi.org/10.1039/D0RA06014C

Kumagai S, Fujita K, Kameda T, Yoshioka T (2016) Interactions of beech wood polyethylene mixtures during co-pyrolysis. J Anal Appl Pyrolysis 122:531–540. ISSN 0165-2370. https://doi.org/10.1016/j.jaap.2016.08.012. https://www.sciencedirect.com/science/article/pii/S0165237016303151

Li Y, Gupta R, You S (2022) Machine learning assisted prediction of biochar yield and composition via pyrolysis of biomass. Bioresour Technol 359:127511. ISSN 0960-8524. https://doi.org/10.1016/j.biortech.2022.127511. https://www.sciencedirect.com/science/article/pii/S0960852422008409

Luo Z, Zhu X, Deng J, Gong K, Zhu X (2021) High-value utilization of mask and heavy fraction of bio-oil: From hazardous waste to biochar, bio-oil, and graphene films. J Hazard Mater 420:126570. ISSN 0304-3894. https://doi.org/10.1016/j.jhazmat.2021.126570. https://www.sciencedirect.com/science/article/pii/S0304389421015351

Nardella F, Bellavia S, Mattonai M, Ribechini E (2022) Co-pyrolysis of biomass and plastic: synergistic effects and estimation of elemental composition of pyrolysis oil by analytical pyrolysis gas chromatography/mass spectrometry. Bioresour Technol 354:127170. ISSN 0960-8524. https://doi.org/10.1016/j.biortech.2022.127170. https://www.sciencedirect.com/science/article/pii/S0960852422004990

Onsree T, Tippayawong N (2021) Machine learning application to predict yields of solid products from biomass torrefaction. Renew Energy 167:425–432

Onsree T, Tippayawong N, Phithakkitnukoon S, Lauterbach J (2022) Interpretable machine-learning model with a collaborative game approach to predict yields and higher heating value of torrefied biomass. Energy 249:123676

Onsree T, Tippayawong N, Williams T, McCullough K, Barrow E, Pogaku R, Lauterbach J (2019) Torrefaction of pelletized corn residues with wet flue gas. Bioresour Technol 285:121330. ISSN 0960-8524. https://doi.org/10.1016/j.biortech.2019.121330. https://www.sciencedirect.com/science/article/pii/S0960852419305528

Patil V, Adhikari S, Cross P (2018) Co-pyrolysis of lignin and plastics using red clay as catalyst in a micro-pyrolyzer. Bioresour Technol 270:311–319. ISSN 0960-8524. https://doi.org/10.1016/j.biortech.2018.09.034. https://www.sciencedirect.com/science/article/pii/S0960852418312872

Phromphithak S, Onsree T, Tippayawong N (2021) Machine learning prediction of cellulose-rich materials from biomass pretreatment with ionic liquid solvents. Bioresour Technol 323:124642. ISSN 0960-8524. https://doi.org/10.1016/j.biortech.2020.124642. https://www.sciencedirect.com/science/article/pii/S0960852420319167

Prasertpong P, Onsree T, Khuenkaeo N, Tippayawong N, Lauterbach J (2023) Exposing and understanding synergistic effects in co-pyrolysis of biomass and plastic waste via machine learning. Bioresour Technol 369:128419. ISSN 0960-8524. https://doi.org/10.1016/j.biortech.2022.128419. https://www.sciencedirect.com/science/article/pii/S0960852422017527

Saengsuriwong R, Onsree T, Phromphithak S, Tippayawong N (2021) Biocrude oil production via hydrothermal liquefaction of food waste in a simplified high-throughput reactor. Bioresour Technol 341:125750. ISSN 0960-8524. https://doi.org/10.1016/j.biortech.2021.125750. https://www.sciencedirect.com/science/article/pii/S0960852421010919

Sanahuja-Parejo O, Veses A, Navarro M, Lopez J, Murillo R, Callen M, Garcia T (2019) Drop-in biofuels from the co-pyrolysis of grape seeds and polystyrene. Chem Eng J 377:120246. ISSN 1385-8947. https://doi.org/10.1016/j.cej.2018.10.183. https://www.sciencedirect.com/science/article/pii/S1385894718321454. ISCRE 25 Special Issue: Bridging Science and Technology

Shapley LS et al (1953) A value for n-person games

Sophonrat N, Sandstrom L, Zaini IN, Yang W (2018) Stepwise pyrolysis of mixed plastics and paper for separation of oxygenated and hydrocarbon condensates. Appl Energy 229:314–325. ISSN 0306-2619. https://doi.org/10.1016/j.apenergy.2018.08.006. https://www.sciencedirect.com/science/article/pii/S0306261918311632

Suriapparao DV, Gautam R, Jeeru LR (2022a) Analysis of pyrolysis index and reaction mechanism in microwave-assisted ex-situ catalytic co-pyrolysis of agro-residual and plastic wastes. Bioresour Technol 357:127357. ISSN 0960-8524. https://doi.org/10.1016/j.biortech.2022.127357. https://www.sciencedirect.com/science/article/pii/S0960852422006861

Suriapparao DV, Sridevi V, Ramesh P, Sankar Rao C, Tukarambai M, Kamireddi D, Gautam R, Dharaskar SA, Pritam K (2022b) Synthesis of sustainable chemicals from waste tea powder and polystyrene via microwave-assisted in-situ catalytic co-pyrolysis: analysis of pyrolysis using experimental and modeling approaches. Bioresour Technol 362:127813. ISSN 0960-8524. https://doi.org/10.1016/j.biortech.2022.127813. https://www.sciencedirect.com/science/article/pii/S0960852422011439

Tang Q, Chen Y, Yang H, Liu M, Xiao H, Wang S, Chen H, Naqvi SR (2021) Machine learning prediction of pyrolytic gas yield and compositions with feature reduction methods: effects of pyrolysis conditions and biomass characteristics. Bioresour Technol 339:125581. ISSN 0960-8524. https://doi.org/10.1016/j.biortech.2021.125581. https://www.sciencedirect.com/science/article/pii/S0960852421009226

Tang Y, Huang Q, Sun K, Chi Y, Yan J (2018) Co-pyrolysis characteristics and kinetic analysis of organic food waste and plastic. Bioresour Technol 249:16–23. ISSN 0960-8524. https://doi.org/10.1016/j.biortech.2017.09.210. https://www.sciencedirect.com/science/article/pii/S0960852417317893

Ullah Z, khan M, Naqvi SR, Farooq W, Yang H, Wang S, Vo D-VN (2021) A comparative study of machine learning methods for bio-oil yield prediction—a genetic algorithm-based features selection. Bioresour Technol 335:125292. ISSN 0960-8524. https://doi.org/10.1016/j.biortech.2021.125292. https://www.sciencedirect.com/science/article/pii/S0960852421006313

Wang F, Harindintwali JD, Yuan Z, Wang M, Wang F, Li S, Yin Z, Huang L, Fu Y, Li L, Chang SX, Zhang L, Rinklebe J, Yuan Z, Zhu Q, Xiang L, Tsang DC, Xu L, Jiang X, Liu J, Wei N, Kästner M, Zou Y, Ok YS, Shen J, Peng D, Zhang W, Barceló D, Zhou Y, Bai Z, Li B, Zhang B, Wei K, Cao H, Tan Z, bin Zhao L, He X, Zheng J, Bolan N, Liu X, Huang C, Dietmann S, Luo M, Sun N, Gong J, Gong Y, Brahushi F, Zhang T, Xiao C, Li X, Chen W, Jiao N, Lehmann J, Zhu Y-G, Jin H, Schäffer A, Tiedje JM, Chen JM (2021a) Technologies and perspectives for achieving carbon neutrality. Innov 2(4):100180. ISSN 2666-6758. https://doi.org/10.1016/j.xinn.2021.100180. https://www.sciencedirect.com/science/article/pii/S2666675821001053

Wang J, Jiang J, Sun Y, Meng X, Wang X, Ruan R, Ragauskas AJ, Tsang DC (2021c) Heterogeneous Diels–Alder tandem catalysis for converting cellulose and polyethylene into BTX. J Hazard Mater 414:125418. ISSN 0304-3894. https://doi.org/10.1016/j.jhazmat.2021.125418. https://www.sciencedirect.com/science/article/pii/S0304389421003812

Wang F, Li W, Bao W, Lv Z (2021b) Gauss–seidel method with oblique direction. Results Appl Math 12:100180. ISSN 2590-0374. https://doi.org/10.1016/j.rinam.2021.100180. https://www.sciencedirect.com/science/article/pii/S2590037421000327

Wang Y, Li Y, Zhang C, Yang L, Fan X, Chu L (2021d) A study on co-pyrolysis mechanisms of biomass and polyethylene via ReaxFF molecular dynamic simulation and density functional theory. Process Saf Environ Prot 150:22–35. ISSN 0957-5820. https://doi.org/10.1016/j.psep.2021.04.002. https://www.sciencedirect.com/science/article/pii/S0957582021001786

Wei H, Luo K, Xing J, Fan J (2022) Predicting co-pyrolysis of coal and biomass using machine learning approaches. Fuel 310:122248. ISSN 0016-2361. https://doi.org/10.1016/j.fuel.2021.122248. https://www.sciencedirect.com/science/article/pii/S0016236121021220

Wongchai W, Onsree T, Sukkam N, Promwungkwa A, Tippayawong N (2022) Machine learning models for estimating above ground biomass of fast growing trees. Expert Syst Appl 199:117186. ISSN 0957-4174. https://doi.org/10.1016/j.eswa.2022.117186. https://www.sciencedirect.com/science/article/pii/S0957417422005735

Yiin CL, Yap KL, Ku AZE, Chin BLF, Lock SSM, Cheah KW, Loy ACM, Chan YH (2021) Recent advances in green solvents for lignocellulosic biomass pretreatment: potential of choline chloride (ChCl) based solvents. Bioresour Technol 333:125195. ISSN 0960-8524. https://doi.org/10.1016/j.biortech.2021.125195. https://www.sciencedirect.com/science/article/pii/S0960852421005344

Zhu X, He M, Sun Y, Xu Z, Wan Z, Hou D, Alessi DS, Tsang DC (2022a) Insights into the adsorption of pharmaceuticals and personal care products (PPCPs) on biochar and activated carbon with the aid of machine learning. J Hazard Mater 423:127060. ISSN 0304-3894. https://doi.org/10.1016/j.jhazmat.2021.127060. https://www.sciencedirect.com/science/article/pii/S0304389421020288

Zhu X, Labianca C, He M, Luo Z, Wu C, You S, Tsang DC (2022b) Life-cycle assessment of pyrolysis processes for sustainable production of biochar from agro-residues. Bioresour Technol 360:127601. ISSN 0960-8524. https://doi.org/10.1016/j.biortech.2022.127601. https://www.sciencedirect.com/science/article/pii/S0960852422009300

Zhu X, Xu Z, You S, Komarek M, Alessi DS, Yuan X, Palansooriya KN, Ok YS, Tsang DC (2022c) Machine learning exploration of the direct and indirect roles of Fe impregnation on Cr(VI) removal by engineered biochar. Chem Eng J 428:131967. ISSN 1385-8947. https://doi.org/10.1016/j.cej.2021.131967. https://www.sciencedirect.com/science/article/pii/S1385894721035476

Chapter 6
Biodiesel Production

Abstract The increasing need for energy has led scientists and engineers to explore renewable sources, and biodiesel has emerged as a promising alternative that is also environmentally friendly. This chapter will look at how machine learning, a type of artificial intelligence, can help us predict how much biodiesel we can get from a process called transesterification, which involves different catalysts. These catalysts can be homogeneous, heterogeneous, or enzyme-based. Of the methods tested, a technique called extreme gradient boosting gave the most accurate predictions. It was able to predict biodiesel yields with a high degree of accuracy, as shown by a determination coefficient of almost 0.98, after being tested rigorously through tenfold cross-validation. The analysis also showed which factors are most important for getting good biodiesel yields. For homogeneous catalysts, linoleic acid levels were critical. For heterogeneous catalysts, it was behenic acid. And for enzyme catalysts, the reaction time was the most important factor. These findings help us understand how these factors, both individually and in combination, affect how well different catalysts work. This understanding is valuable for making biodiesel production more efficient.

Keywords Artificial intelligence · Extreme gradient boosting · Renewable energy · Biofuel · Transesterification catalysts

6.1 Introduction

Over the last decade, there has been increasing awareness of the need for energy sources that are cleaner, safer, and more sustainable. Biodiesel has become a popular alternative to conventional diesel fuel, which comes from limited fossil fuel sources. With the world population expected to exceed 9 billion by 2050 and energy use predicted to rise significantly, it's crucial to find alternative energy solutions to tackle current and future energy problems (Kumar et al. 2020; Zhao et al. 2021).

Several factors are driving the shift toward biodiesel, including the fact that fossil fuels are finite, petroleum prices increase, and traditional diesel has negative environmental effects, such as air pollution (Modiba et al. 2014). Biodiesel offers a cleaner,

N. Tippayawong et al., *Machine Learning Applications for Bioenergy Conversion*, Green Energy and Technology, https://doi.org/10.1007/978-981-96-8086-3_6

more environmentally friendly option, producing fewer emissions of carbon monoxide and particulate matter. It also lacks sulfur and aromatic compounds, ensuring cleaner combustion (Krishnamurthy et al. 2018). Biodiesel reduces air pollution and plays a vital role in moving toward a more sustainable energy future, especially in transportation and energy production.

Biodiesel production involves a chemical process that converts free fatty acids (FFAs), oils, or fats into fatty acid methyl esters (FAMEs) using either esterification or transesterification reactions (Tacias-Pascacio et al. 2019).

1. Transesterification: The Primary Method Transesterification is the main reaction for converting triglycerides (oils/fats) into biodiesel (FAMEs).

- Process: Alcohol (e.g., methanol) reacts with triglycerides in the presence of a base catalyst (NaOH, KOH) to produce biodiesel and glycerol.
- Conditions:
  - Base-catalyzed: Requires low FFA ($<0.5\%$) and anhydrous conditions to avoid soap formation.
  - Efficiency: Achieves $>95\%$ yield under optimal conditions (60–70 °C, 1–2 h, methanol-to-oil molar ratio ~6:1).

2. Esterification: A Pre-treatment for High-FFA Feedstocks Esterification is used as a pre-treatment step for feedstocks with high free fatty acid (FFA) content (>1–2%), such as waste cooking oil or animal fats.

- Process: Acid catalysts (e.g., H_2SO_4) esterify FFAs into esters, reducing FFA levels to $<1\%$.
- Example:
  - Using 3% H_2SO_4, 1:6 oil-to-methanol ratio at 50 °C reduces FFA in used cooking oil from 4.26% to 2.11%.
- Purpose: Prevents soap formation during subsequent base-catalyzed transesterification

Integrated Process for Low-Quality Feedstocks For feedstocks with high FFA (e.g., used cooking oil):

1. Esterification: Reduces FFA content to avoid saponification. Saponification is the chemical process of turning fats or oils into soap and glycerol (glycerin) by reacting them with a strong base.
2. Transesterification: Converts pre-treated oil into biodiesel.

- Example: After esterification with H_2SO_4, transesterification with NaOH yields biodiesel.

When Are They Used Separately?

- Transesterification alone: Suitable for refined oils with low FFA (e.g., soybean oil).

- Esterification alone: Rarely used for full biodiesel production; primarily a pretreatment step.

Catalysts—classified as homogeneous, heterogeneous, or enzyme-based—help these reactions occur, and their selection depends on things like feedstock type, operating conditions, desired efficiency, cost, and availability (Jayakumar et al. 2021; Endalew et al. 2011). Homogeneous catalysts dissolve in the reaction mixture, while heterogeneous catalysts remain separate. They both can be tailored to specific applications, improving the oil-to-biodiesel conversion process. Batch reaction methods are commonly used to maximize FAME yields. The choice of catalyst is crucial for determining biodiesel production efficiency and yield (Almohana et al. 2022). A brief overview of each catalyst is shown in Table 6.1.

As biodiesel production gains attention, researchers are studying the factors affecting its yield. These factors fall into two categories: controllable and uncontrollable. Controllable factors include variables like the oil-to-alcohol ratio, catalyst type and amount, reaction temperature, and time (Rashid et al. 2011). Their impact has been extensively studied (Mathiyazhagan and Ganapathi 2011). On the other hand, uncontrollable factors involve feedstock characteristics, such as the fatty acid content (e.g., palmitic acid, stearic acid, oleic acid, linoleic acid) (Pinzi et al. 2010). Many studies, however, focus narrowly on specific feedstocks or catalysts, limiting our understanding of the complex interactions among all variables involved in biodiesel production. Addressing these gaps remains a challenge, but ongoing research continues to explore ways to optimize biodiesel production processes.

The fatty acid composition of feedstocks significantly impacts biodiesel yield and quality through several key mechanisms:

1. Saturated Versus Unsaturated Fatty Acids

- Saturated fatty acids (e.g., palmitic, stearic) improve oxidation stability but worsen cold flow properties, making biodiesel less suitable for colder climates.
- Unsaturated fatty acids (e.g., oleic, linoleic) enhance cold flow but reduce oxidation stability, increasing the risk of degradation during storage.

Table 6.1 Comparison table for homogeneous, heterogeneous and enzyme-based catalysts

Property	Homogeneous	Heterogeneous	Enzyme-based
Phase	Liquid	Solid	Solid/Liquid (immobilized)
FFA tolerance	Low (<2%)	Moderate (up to 5%)	High (any FFA)
Reusability	No	Yes	Yes (with immobilization)
Reaction speed	Fast (1–2 h)	Moderate (4–10 h)	Slow (6–24 h)
Cost	Low	Moderate	High
Byproducts	Soap, wastewater	Minimal	Glycerol (easily separable)
Industrial use	Common (refined oils)	Emerging (waste oils)	Niche (high-value applications)

- Feedstocks rich in monounsaturated fats (e.g., oleic acid in Jatropha curcas) strike a balance, offering better oxidative stability than polyunsaturated fats while maintaining reasonable cold flow properties.

2. Free Fatty Acid (FFA) Content

- High-FFA levels in feedstocks (e.g., low-quality oils) can lead to soap formation during transesterification, reducing biodiesel yield and complicating purification.
- Pretreatment (e.g., acid esterification) is often required for feedstocks with FFA $> 1\%$ to avoid yield losses.

3. Fatty Acid Chain Length

- Longer carbon chains (e.g., lignoceric acid) increase viscosity, potentially affecting engine performance, whereas shorter chains improve flow properties.
- Optimal chain lengths (C16–C18) are preferred for balancing fuel properties and yield.

4. Feedstock Blending

- Combining oils with complementary fatty acid profiles (e.g., high saturation + high unsaturation) can create composite biodiesel with improved overall properties, reducing reliance on single feedstocks.

5. Regional and Environmental Variability

- Fatty acid composition varies with geographic origin due to soil and climate, directly affecting biodiesel quality and yield consistency.

Practical Implications

- High-oleic feedstocks (e.g., soybean, sunflower) are favored for their balanced properties.
- Polyunsaturated-rich oils (e.g., linseed) may require additives to stabilize oxidation.
- Blending or genetic modification of crops can optimize fatty acid profiles for higher yields and better fuel performance.

Machine learning (ML) provides powerful tools for understanding complicated information, identifying patterns, and improving biodiesel production. This chapter explores how ML can be used to predict biodiesel purity, focusing on which factors are most important and how they affect the accuracy of predictions. Using a collection of 270 data points with eight different factors, the study assesses the validity of these factors using statistical techniques like correlation analysis and feature selection algorithms. The findings offer insights into the most relevant factors for predicting biodiesel purity and guide the development of an optimized ML model. These results contribute to advancements in biodiesel production and process optimization through ML applications (Gupta et al. 2021).

One model, called an artificial neural network (ANN), stands out for its high accuracy in predicting diesel engine performance and exhaust emissions using biodiesel fuels, achieving a determination coefficient above 95%. Other studies show how ML can increase biodiesel yield, reduce CO_2 emissions, and lower costs. A study combined ANN modeling and response surface methodology (RSM) to optimize waste-derived biogas and algal biodiesel as alternative fuels in dual-fuel engines (Sharma et al. 2023). These examples highlight ML's potential to transform biodiesel-related processes and decision-making.

In this chapter, an ML model is developed to predict biodiesel yield percentage (%BDY) based on process parameters, aiming to quantitatively assess how these variables influence production. The focus is on understanding relationships between process parameters and response characteristics while building a predictive model to optimize biodiesel production. By employing algorithms such as Support Vector Regression (SVR), Random Forest (RF), Extreme Gradient Boosting (XGB), and Kernel Ridge Regression (KRR), the study provides actionable insights into the transesterification process. One-dimensional and two-dimensional Partial Dependence Plots (1D-PDPs and 2D-PDPs) are used to visualize the impact of individual features on %BDY.

The research also investigates the role of three catalyst types in biodiesel production, analyzing their effects on %BDY predictions. The results offer a deeper understanding of how operating conditions and process features influence biodiesel yields. These findings highlight ML's ability to identify key factors, improve process efficiency, and guide decision-making in biodiesel production. By leveraging ML, organizations can handle large datasets, automate repetitive tasks, and achieve more accurate and efficient outcomes, ultimately contributing to a more sustainable energy future.

6.2 Methodology

Successfully applying machine learning (ML) to predict biodiesel yield percentage (%BDY) relies on a structured, well-organized dataset. This chapter uses data gathered from published studies to develop and test ML models, focusing on 144 distinct feedstocks like castor oil, soybean oil, and jatropha oil. These feedstocks highlight how different input variables influence biodiesel production outcomes.

The dataset is divided into two key parts. The first part details the fatty acid composition of each feedstock, expressed as weight percentages. This includes 21 fatty acids, such as capric acid (Cap), lauric acid (Lau), myristic acid (Myr), myristolenic acid (Myi), pentadecanoic acid (Pen), palmitic acid (Pal), palmitoleic acid (Pam), margaric acid (Mar), heptadecanoic acid (Hep), stearic acid (Ste), oleic acid (Ole), ricinoleic acid (Ric), linoleic acid (Lin), linolenic acid (Lio), arachidic acid (Ara), eicosenoic acid (Eic), arachidonic acid (Arc), behenic acid (Beh), erucic acid (Eru), lignoceric acid (Lig), and nervonic acid (Ner). These components play a critical role in determining the efficiency of biodiesel production, as fatty acid profiles affect chemical reactions during processing.

The second part focuses on operational parameters that impact the transesterification reaction—the chemical process that converts oils into biodiesel. Key variables include:

- Molar ratio of alcohol to oil (MR): Influences reaction completeness.
- Reaction time (RT): Duration of the process in minutes.
- Reaction temperature (Temp): Measured in degrees Celsius.
- Catalyst concentration (Cat): Expressed as a weight percentage.

By systematically capturing these factors, the dataset provides a robust foundation for training ML models to predict %BDY accurately. This structured approach ensures that both feedstock characteristics and process conditions are accounted for, enabling models to identify patterns and optimize production.

Studies like Gupta et al. (2021) and Sharma et al. (2023) demonstrate how such datasets enable ML techniques—including artificial neural networks (ANNs) and extreme gradient boosting—to achieve high prediction accuracy (e.g., determination coefficients above 0.95). This chapter underscores the importance of comprehensive data collection in advancing biodiesel research, offering insights into how ML can streamline production processes and support sustainable energy goals.

To gather the necessary data for this study, an extensive literature search was conducted using Scopus, Web of Science, and Google Scholar. Keywords such as "biodiesel," "transesterification," "response surface," "Central Composite Design," and "Box-Behnken design" were employed to ensure the retrieval of relevant and useful data. A Central Composite Design (CCD) is a widely used experimental design technique in response surface methodology, allowing for efficient modeling of a quadratic relationship between variables without the need for a full factorial experiment. A Box-Behnken design (BBD) is a response surface methodology used in statistical experiments to fit a quadratic model, efficiently exploring the experimental space and avoiding extreme conditions. This strategic combination of search engines and keywords enabled a thorough and comprehensive data collection process.

For predicting %BDY, the transesterification process data were categorized according to the type of catalyst used, resulting in three distinct categories: homogeneous catalysts, heterogeneous catalysts, and enzymatic catalysts.

- Homogeneous catalysts (Case 1): A total of 404 data points were collected from 51 peer-reviewed publications covering 55 types of feedstock. This dataset included 20 variables, which are detailed in Table 6.2.
- Heterogeneous catalysts (Case 2): A larger dataset of 491 data points was derived from 65 peer-reviewed publications and 69 types of feedstock. This dataset comprised 24 critical features, as presented in Table 6.3.
- Enzymatic catalysts (Case 3): For this category, 113 data points were sourced from 19 peer-reviewed publications involving 24 types of feedstock. This dataset included 19 key variables, outlined in Table 6.4.

Table 6.2 Set of input features/variables for homogeneous catalysts in case 1, from Sukpancharoen et al. (2023) (with permission)

Features	Formulation	Name	Value	Mean	SD
Fatty acids composition (wt. %)					
Capric acid	C10:0	Cap	0–3.40	0.23	0.67
Lauric acid	C12:0	Lau	0–48.20	0.44	3.43
Myristic acid	C14:0	Myr	0–16.20	1.10	3.19
Myristolenic acid	C14:1	Myi	0–0.20	0.001	0.01
Palmitic acid	C16:0	Pal	0–49.80	15.92	11.68
Palmitoleic acid	C16:1	Pam	0–4.96	2.90	0.73
Stearic acid	C18:0	Ste	0–24.50	6.13	4.41
Oleic acid	C18:1	Ole	3.53–80.57	32.20	18.86
Ricinoleic acid	C18:1(OH)	Ric	0–89.27	7.33	23.31
Linoleic acid	C18:2	Lin	0.67–77.07	26.68	21.14
Linolenic acid	C18:3	Lio	0–51.79	6.26	14.00
Arachidic acid	C20:0	Ara	0–27.21	0.30	1.45
Eicosenoic acid	C20:1	Eic	0–2.00	0.14	0.41
Behenic acid	C22:0	Beh	1–2.80	0.14	0.45
Erucic acid	C22:1	Eru	0–1.51	0.06	0.29
Lignoceric acid	C24:0	Lig	0–0.99	0.04	0.20
Operating condition					
Molar ratio of alcohol to oil (mol/mol)		MR	1.60–12.55	6.93	2.36
Reaction time (mins)		RT	0.50–255.00	52.44	34.40
Reaction temperature (°C)		Temp	24.80–90.00	54.13	12.60
Catalyst concentration (wt. %)		Cat	0–7.50	1.11	0.78
Output target					
Biodiesel yield (%)		% BDY	60.00–99.75	86.87	8.19

To prepare the data for analysis, all input variables were normalized using the Yeo-Johnson transformation, as described in Sect. 2.7. This normalization process ensured the datasets were suitable for use in developing accurate machine learning models.

6.2.1 Machine Learning Model Development

This chapter employs advanced machine learning algorithms—Support Vector Regression (SVR), Random Forest (RF), Extreme Gradient Boosting (XGB), and Kernel Ridge Regression (KRR)—to predict biodiesel yield percentage (%BDY). These algorithms were implemented using Python's Scikit-learn library, a widely

Table 6.3 Set of input features/variables for heterogeneous catalysts in case 2, from Sukpancharoen et al. (2023) (with permission)

Features	Formulation	Name	Value	Mean	SD
Fatty acids composition (wt. %)					
Capric acid	C10:0	Cap	0–4.82	0.18	0.60
Lauric acid	C12:0	Lau	0–59.83	4.08	13.16
Myristic acid	C14:0	Myr	0–15.65	1.64	4.32
Myristolenic acid	C14:1	Myi	0–2.00	0.01	0.13
Pentadecanoic acid	C15:0	Pen	0–1.90	0.01	0.12
Palmitic acid	C16:0	Pal	2.20–53.01	18.21	12.07
Palmitoleic acid	C16:1	Pam	0–6.30	0.79	1.62
Margaric acid	C17:0	Mar	0–1.62	0.06	0.30
Heptadecanoic acid	C17:1	Hep	0–0.08	0.00	0.00
Stearic acid	C18:0	Ste	0–30.94	8.32	7.90
Oleic acid	C18:1	Ole	6.87–82.05	39.45	13.05
Linoleic acid	C18:2	Lin	0–63.50	21.86	16.62
Linolenic acid	C18:3	Lio	0–54.00	2.97	8.48
Arachidic acid	C20:0	Ara	0–26.44	0.83	3.66
Eicosenoic acid	C20:1	Eic	0–14.80	0.39	2.07
Arachidonic acid	C20:4	Arc	0–1.30	0.03	0.15
Behenic acid	C22:0	Beh	0–5.70	0.21	0.68
Erucic acid	C22:1	Eru	0–37.78	0.09	1.72
Lignoceric acid	C24:0	Lig	0–1.40	0.06	0.24
Nervonic acid	C24:1	Ner	0–4.10	0.01	0.19
Operating condition					
Molar ratio of alcohol to oil (mol/mol)		MR	1–82	11.64	11.27
Reaction time (mins)		RT	15–480	128.92	94.76
Reaction temperature (°C)		Temp	25–210	72.32	27.77
Catalyst concentration (wt. %)		Cat	0.1–100	4.75	5.75
Output target					
Biodiesel yield (%)		% BDY	45.60–99.90	83.75	12.44

used tool for machine learning tasks (Katongtung et al. 2022). To handle the computational demands of these models, the study used a high-performance computer system with a 12-core processor and 32 GB of memory, enabling efficient processing of complex calculations within practical time frames.

To ensure reliable results, the models were tested using a tenfold cross-validation method. This technique divides the dataset into 10 equal parts, using nine parts to train the model and one part to test its performance. This process repeats 10 times, cycling through each subset as the test data, and the results are averaged to provide a robust evaluation of how well the models generalize to new, unseen data. This

Table 6.4 Set of input features/variables for enzymatic catalysts in case 3, from Sukpancharoen et al. (2023) (with permission)

Features	Formulat-ion	Name	Value	Mean	SD
Fatty acids composition (wt. %)					
Capric acid	C10:0	Cap	0–2.70	0.36	0.92
Lauric acid	C12:0	Lau	0–1.20	0.08	0.23
Myristic acid	C14:0	Myr	0–3.90	0.10	0.45
Palmitic acid	C16:0	Pal	0–43.50	12.71	8.23
Palmitoleic acid	C16:1	Pam	0–16.25	0.94	1.90
Margaric acid	C17:0	Mar	0–0.07	0.02	0.03
Heptadecanoic acid	C17:1	Hep	0–0.15	0.04	0.07
Stearic acid	C18:0	Ste	1.30–30.00	3.57	3.63
Oleic acid	C18:1	Ole	4.50–80.46	39.80	25.41
Ricinoleic acid	C18:1 (OH)	Ric	0–88.20	18.26	35.56
Linoleic acid	C18:2	Lin	0–66.52	16.28	13.37
Linolenic acid	C18:3	Lio	0–44.33	6.75	14.79
Arachidic acid	C20:0	Ara	0–0.40	0.10	0.15
Eicosenoic acid	C20:1	Eic	0–0.10	0.02	0.04
Behenic acid	C22:0	Beh	0–0.16	0.04	0.07
Operating condition					
Molar ratio of alcohol to oil (mol/mol)		MR	1–15	5.62	2.37
Reaction time (mins)		RT	30–4320	841.86	701.36
Reaction temperature (°C)		Temp	30–180	42.54	15.84
Catalyst concentration (wt. %)		Cat	0–40	8.14	5.60
Output target					
Biodiesel yield (%)		% BDY	35.60–99.00	79.23	14.88

approach helps mitigate over fitting and provides a robust assessment of the model's ability to generalize to new data (Vu et al. 2022).

Three key metrics were used to evaluate model accuracy:

1. R2 (coefficient of determination): Measures how closely predicted values match actual data, with 100% indicating perfect alignment.
2. Normalized Mean Square Error (NMSE): Quantifies the average squared difference between predictions and observations, scaled for comparison across datasets.
3. Normalized Mean Absolute Error (NMAE): Reflects the average absolute difference between predictions and real values, normalized for consistency.

These metrics are widely used in ML model evaluations and are discussed in detail in Sect. 2.7.

Table 6.5 Optimized hyperparameters of the four algorithms and three cases data set to predict %BDY from Sukpancharoen et al. (2023) (with permission)

ML algorithm	Hyperparameter estimator	Optimized value		
		Case 1	Case 2	Case 3
SVR	C	10	10	10
	Kernel	rbf	rbf	rbf
	Degree	2	2	2
	Gamma	Scale	Scale	Scale
	Tol	0.001	0.001	0.001
	Shrinking	TRUE	TRUE	TRUE
RF	Max. depth	5	5	5
	Min. samples split	1	1	1
	Min. sample leaf	5	5	5
	Warm start	FALSE	FALSE	FALSE
XGB	Eta	0.5	0.7	0.5
	Gamma	0	0	0
	Min. child weight	0	0	0
	Max. depth	9	6	6
	Max. depth step	0	0	0
KRR	Alpha	10	1	10
	Degree	4	4	4
	Gamma	10	10	20
	Kernel	Polynomial	Polynomial	Polynomial

6.3 Machine Learning Algorithm

To identify the algorithm with the best predictive performance, each was customized using the full-factor grid search method to optimize hyperparameters (Qu et al. 2021; Shi et al. 2022). Table 6.5 presents the optimized hyperparameters for each algorithm and dataset.

The four machine learning algorithms—Support Vector Regression (SVR), Random Forest (RF), Extreme Gradient Boosting (XGB), and Kernel Ridge Regression (KRR)—each employ distinct approaches to analyze data patterns and improve prediction accuracy, as explained in detail in Chap. 2.

- Support Vector Regression (SVR) identifies relationships in complex datasets by focusing on key data points called support vectors. It creates an optimal boundary (hyperplane) to predict outcomes, making it effective for non-linear patterns (Kang et al. 2023).
- Random Forest (RF) combines multiple decision trees to enhance prediction reliability. By analyzing different subsets of data features, RF reduces over fitting risks and improves accuracy compared to single-tree models.

- Extreme Gradient Boosting (XGB) sequentially builds decision trees, with each subsequent model addressing errors from previous iterations. This step wise refinement makes it particularly effective for complex datasets.
- Kernel Ridge Regression (KRR) handles non-linear relationships by transforming data into higher dimensions while maintaining computational efficiency. Unlike SVR, KRR uses squared error calculations with regularization to balance accuracy and model simplicity.

The combination of these algorithms allows researchers to address different aspects of biodiesel yield prediction—from handling complex variable interactions (via RF and XGB) to modeling non-linear relationships (through SVR and KRR). This versatility is crucial given the diverse factors affecting biodiesel production, including catalyst types, reaction times, and fatty acid profiles in feedstocks. As shown in recent multi-input models, such approaches can simultaneously optimize yield percentages (up to 96%) while maintaining critical fuel properties.

6.3.1 Feature Evaluation and Interpretation

To evaluate the influence of input features on prediction accuracy across the four machine learning (ML) algorithms—SVR, RF, XGB, and KRR—an experimental study was conducted. The study involved training and testing the models on three distinct datasets:

1. Case 1: Homogeneous catalysts.
2. Case 2: Heterogeneous catalysts.
3. Case 3: Enzymatic catalysts.

The hyperparameters for each ML algorithm were optimized using a full grid search approach tailored to the datasets. Additionally, feature importance rankings were derived from the inbuilt feature selection mechanisms of each algorithm, identifying the features most critical to predicting %BDY (Phromphithak et al. 2021).

To further analyze catalyst behavior during the transesterification process, three highly influential features common to all three cases were selected for deeper exploration. These features underwent 1D Partial Dependence Plot (PDP) analysis to reveal their primary relationship with %BDY. Additionally, 2D-PDP analysis was applied to examine the interaction between feature pairs and their collective impact on %BDY (Zhao et al. 2022), as shown in Eq. 6.1:

$$\hat{F}(X_s) = \frac{1}{n}[1]n\sum \hat{F}(X_s, X_c^i) \tag{6.1}$$

where X_s is the set of features of interest (1–2 features), X_c is a set of features other than those of interest and $\hat{F}$ represents a trained regression model that produces a predicted value.

6.4 Results and Discussion

Pearson analysis To explore potential feature overlaps and enhance predictive modeling for %BDY estimation, the Pearson Correlation Coefficient (PCC) was employed to measure the collinearity between pairs of independent variables. A heat map was used to visualize the linear dependencies within the datasets, with separate heat maps provided for each case:

1. Case 1: Homogeneous catalysts (Fig. 6.1a).
2. Case 2: Heterogeneous catalysts (Fig. 6.1b).
3. Case 3: Enzymatic catalysts (Fig. 6.1c).

Key Observations from PCC Analysis:

- A PCC value of 1 indicates a strong positive correlation, while 0 signifies no correlation.
- Variables with $|PCC| > 0.6$ were deemed correlated, and one of the correlated variables was excluded from model development to prevent redundancy (Hai et al. 2023).
- The sign of the PCC indicates the nature of the relationship:
  - Positive values reflect a direct correlation.
  - Negative values reflect an inverse correlation.

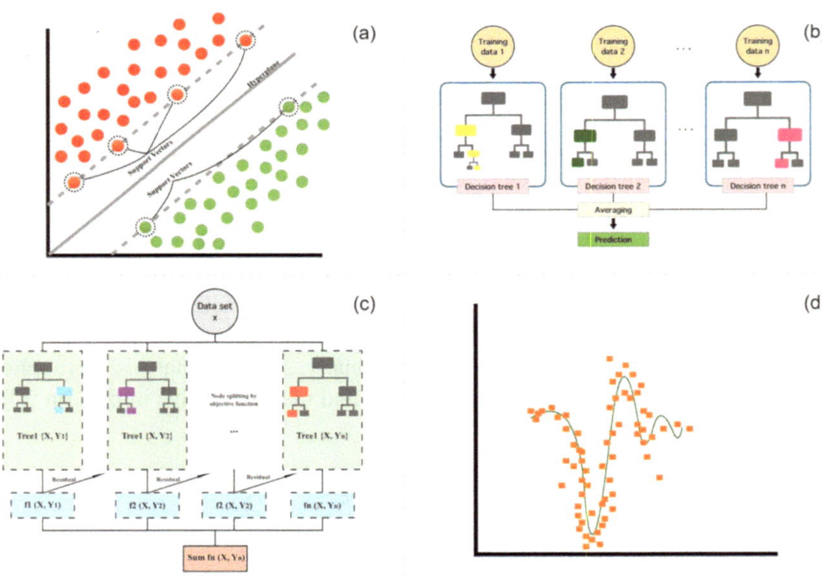

Fig. 6.1 Pearson correlation of data sets for **a** case 1, **b** case 2, and **c** case 3, from Sukpancharoen et al. (2023) (with permission)

Case 1 Observations (Fig. 6.1a):

- Beh and Lig features: |PCC| = 0.88.
- Cap and Myr features: |PCC| = 0.75.
- Ole and Eic features: |PCC| = 0.67.
- Eic and Lig features: |PCC| = 0.67.
- Beh and Eic features: |PCC| = 0.60.

For all other features, the PCC values were below 0.6, indicating low correlation.
Feature Selection Strategy To build reliable machine learning models, this chapter carefully manages relationships between different input variables. The percentage of fatty acid composition was maintained even those with strong correlations. This choice ensures that all fatty acids add up to 100%, reflecting real-world accuracy and consistency.

6.4.1 Model Selection and Accuracy

Table 6.6 shows how well four different machine learning models—SVR, RF, XGB, and KRR—predict biodiesel production outcomes across three different sets of data. Each model was fine-tuned to perform at its best. Among them, the Extreme Gradient Boosting (XGB) model consistently outperformed the others. It achieved the highest accuracy, as measured by R2, with values of 0.977, 0.975, and 0.983 for the three datasets. This indicates that XGB's predictions aligned very closely with the actual results.

Additionally, XGB had the smallest errors compared to the other models. Its Normalized Mean Square Error (NMSE) values were between 0.0008 and 0.0013,

Table 6.6 Cross-validation prediction accuracy (R^2, NMSE, NMAE) of ML algorithms for all three cases from Sukpancharoen et al. (2023) (with permission)

Model	Input	%BDY		
		R^2(SD.)	NMSE (SD.)	NMAE (SD.)
SVR	Case 1	0.7375 (0.0775)	0.0107 (0.0022)	0.0795 (0.0053)
	Case 2	0.7961 (0.0209)	0.0106 (0.0011)	0.0832 (0.0043)
	Case 3	0.7742 (0.1716)	0.0102 (0.0028)	0.0843 (0.0117)
RF	Case 1	0.7075 (0.0434)	0.0121 (0.0013)	0.0862 (0.0047)
	Case 2	0.6411 (0.0530)	0.0186 (0.0018)	0.1046 (0.0053)
	Case 3	0.8104 (0.1091)	0.0090 (0.0037)	0.0700 (0.0121)
XGB	Case 1	0.9770 (0.0063)	0.0010 (0.0002)	0.0134 (0.0012)
	Case 2	0.9750 (0.0097)	0.0013 (0.0005)	0.0148 (0.0033)
	Case 3	0.9831 (0.0160)	0.0008 (0.0006)	0.0105 (0.0047)
KRR	Case 1	0.9637 (0.0422)	0.0015 (0.0016)	0.0143 (0.0036)
	Case 2	0.9064 (0.1143)	0.0048 (0.0055)	0.0307 (0.0077)
	Case 3	0.9825 (0.0163)	0.0008 (0.0006)	0.0108 (0.0046)

and its Normalized Mean Absolute Error (NMAE) values ranged from 0.0105 to 0.0148.

The Kernel Ridge Regression (KRR) model ranked second in performance, followed by Support Vector Regression (SVR) and Random Forest (RF). To visualize these results, Fig. 6.2 compares the predicted values against the actual values for normalized biodiesel yield (%BDY) in each dataset. In these graphs, a black trend line represents perfect predictions, where the predicted and actual values match exactly.

Each case highlights the two top-performing models. For case 1, XGB and KRR are shown in green in Fig. 6.2a and b. For the second case, XGB and KRR are represented in pink in Fig. 6.2c and d. For the third dataset, brown is used to show XGB and KRR in Fig. 6.2e and f. The solid and dotted lines in Fig. 6.2 indicate the range where predictions deviate by no more than 20% from the actual results, for both the training and test datasets.

In conclusion, the findings show that XGB is the most reliable model for predicting biodiesel yield across all three datasets. Its high accuracy (high R2 values) and low error rates make it an excellent choice for precise predictions compared to the other models tested.

6.4.2 Feature Evaluation

This chapter uses the Extreme Gradient Boosting (XGB) model to determine which factors most influence biodiesel production. The XGB model helps to identify the importance of different input variables, such as the fatty acid composition of feedstocks and transesterification conditions. These were analyzed for three types of catalysts: homogeneous, heterogeneous, and enzymatic.

Figure 6.3a, b, and c show the rankings of these features for each catalyst type. The XGB model calculates a feature importance score that accounts for both linear and non-linear dependencies (Panapitiya et al. 2018).

For homogeneous catalysts, linoleic acid (Lin) was found to be the most significant factor, explaining 34.087% of the variations in biodiesel yield. This is because linoleic acid can react with alcohol, which reduces the availability of alcohol for the main transesterification process, thus lowering the biodiesel yield.

For heterogeneous catalysts, behenic acid (Beh) had the greatest impact, accounting for 36.61% of the yield variations. With enzymatic catalysts, reaction time (RT) was the most critical, contributing 43.18% to biodiesel yield. Enzymatic transesterification involves converting triglycerides into biodiesel, where reaction time plays a key role. While longer reaction times typically increase yields (Yucel 2012), overly long duration's may hinder enzyme activity due to the formation of a fatty acid methyl ester (FAME) layer at the oil-water interface.

Transesterification conditions, including catalyst amount (Cat), reaction time (RT), temperature (Temp), and molar ratio (MR), were also analyzed. Catalyst amount had the highest importance score (4.74%) for homogeneous catalysts, consistent with findings from Jeong and Park (2009). In heterogeneous catalysts, reaction temperature had the highest influence, with a feature importance score of 2.48%.

Fig. 6.2 Comparison of predicted and actual %BDY values from ML models for **a** XGB and **b** KRR in case 1, **c** XGB and **d** KRR in case 2, and **e** XGB and **f** KRR in case 3, from Sukpancharoen et al. 2023 (with permission)

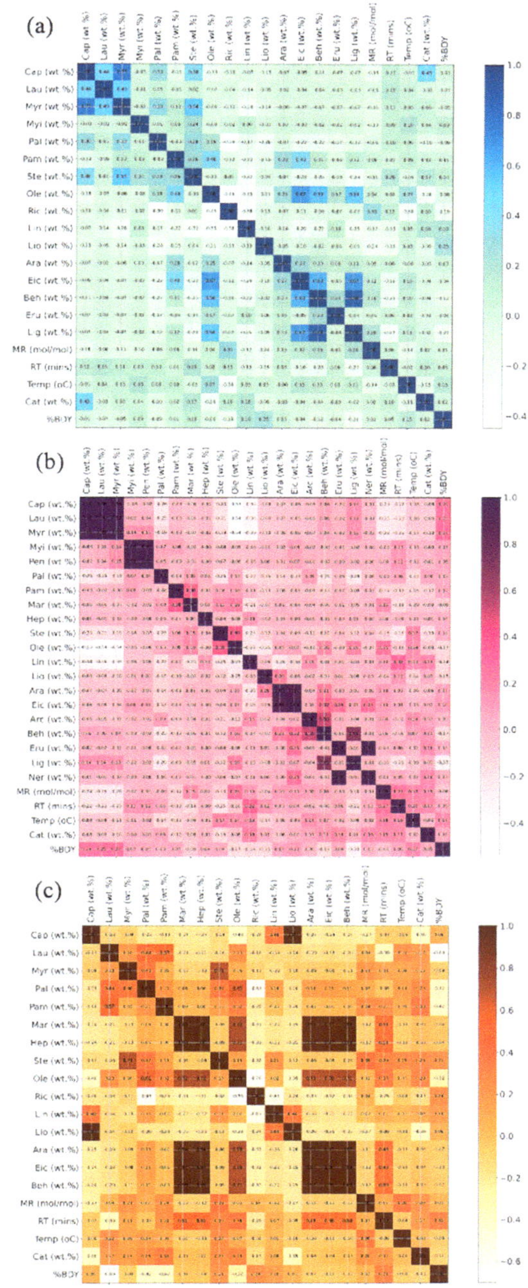

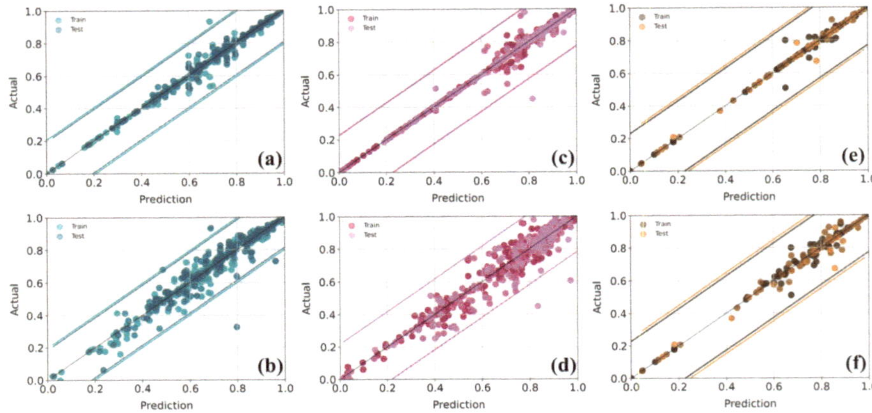

Fig. 6.3 Feature importance of XGB in **a** case 1, **b** case 2 and **c** case 3, from Sukpancharoen et al. (2023) (with permission)

Overall, transesterification conditions had a smaller effect compared to fatty acid composition, indicating that the model's accuracy is less sensitive to variations in these conditions.

The fatty acid composition of the feedstock consistently ranked among the top five features across all cases except enzymatic catalysts, where linolenic acid played a larger role. The presence of linolenic acid, an unsaturated fatty acid with three double bonds, accounted for 32.11% of variations in enzymatic catalysis. Controlling the amount of linolenic acid in raw materials is critical for producing high-quality biodiesel and ensuring stable yields. As prior studies show, XGBoost models can determine key factors for optimizing a biomass gasification system.

In conclusion, both fatty acid composition and transesterification conditions are crucial in biodiesel production. Careful selection and control of raw materials and process parameters are essential for optimal results. With machine learning techniques, the biodiesel yield can be increased and optimized by determining the most important production parameters.

6.4.3 Partial Dependence Plots and Interpretation

This chapter uses 1D-PDP (one-dimensional partial dependence plot) analysis to show how reaction time, temperature, and catalyst concentration affect biodiesel yield (%BDY) during the transesterification process. A partial dependence plot (PDP) helps visualize the average relationship between a feature and the predicted outcome of a machine learning model. The analysis is divided by catalyst type: homogeneous (Fig. 6.4a), heterogeneous (Fig. 6.4b), and enzymatic (Fig. 6.4c).

For homogeneous catalysts, commonly used types include alkali catalysts (e.g., sodium hydroxide, potassium hydroxide) and acidic catalysts (e.g., sulfuric acid, hydrochloric acid) (Ma and Hanna 1999). As shown in Fig. 6.4a, the reaction occurs quickly, with the %BDY reaching 87% within the first 10 min. This rapid yield surpasses the performance of heterogeneous and enzyme catalysts during the same period. However, excessive catalyst concentration can reduce %BDY due to soap formation, which occurs when free fatty acids in the raw oil react with water (Agarwal 2007). Acidic catalysts, in contrast, enhance %BDY by efficiently catalyzing the transesterification reaction, converting triglycerides into alkyl esters and glycerol.

For heterogeneous catalysts, such as metal complexes, hydroxides, and oxides, Fig. 6.4b shows a moderate %BDY of 75–77% at 25–30 °C, with stable reaction rates. These catalysts operate in a solid phase, leading to poor mixing and reactivity compared to homogeneous catalysts in a liquid phase (Zabeti et al. 2009). Raising the temperature above 50 °C improves the reaction rate, increasing %BDY to approximately 86%. However, further temperature increases to 65 °C do not significantly boost the yield. Despite slower reaction rates, heterogeneous catalysts offer practical advantages, such as reducing process steps and eliminating the need for biodiesel purification through water washing.

The effect of reaction time is particularly evident in Fig. 6.4b. For heterogeneous catalysts, extending the reaction time increases %BDY, which peaks around 40 min. After this point, the yield stabilizes, as the reaction becomes limited by mass transfer processes. In heterogeneous systems, the reaction occurs on the catalyst surface, and as the reactant's concentration decreases over time, mass transfer becomes the rate-limiting step. Prolonged reaction times allow reactants to diffuse and complete the reaction, but excessive duration's may trigger reverse reactions, reducing the yield. Optimizing reaction time is therefore essential to balance reaction kinetics and mass transfer, maximizing %BDY without causing unwanted side effects.

In summary, each catalyst type exhibits distinct behaviors in terms of reaction speed, temperature sensitivity, and scalability. Homogeneous catalysts excel in rapid biodiesel production, heterogeneous catalysts simplify downstream processing, and careful time and temperature adjustments can optimize yields for all catalyst types.

Figure 6.4b illustrates how catalyst concentration affects biodiesel yield (%BDY). The amount of catalyst has a minimal effect, as seen by the gentle slope of the curve compared to reaction time and temperature. A catalyst concentration of about 10% by weight resulted in the highest %BDY. Unlike homogeneous catalysts, heterogeneous catalysts don't cause soap formation and are easily separated from biodiesel through filtration. However, excessive catalyst amounts increase disposal costs as noted by Agarwal (2007).

For enzyme-based catalysts, a longer reaction time, often up to 300 min, is needed to achieve the highest %BDY, as shown in Fig. 6.4c. During this time, the reaction proceeds quickly, and small changes in temperature have minimal effect. However, enzymes are protein-based and sensitive to high temperatures, which can lead to denaturation. Enzymes remain functional at temperatures as low as 30 °C, making temperature control essential for biodiesel production using enzymatic catalysts.

The interaction between two operational features on %BDY was further analyzed using 2D-PDPs, a tool designed to identify influential and controllable factors. Figure 6.5a, c, and e show the interaction between MR (molar ratio) and RT (reaction

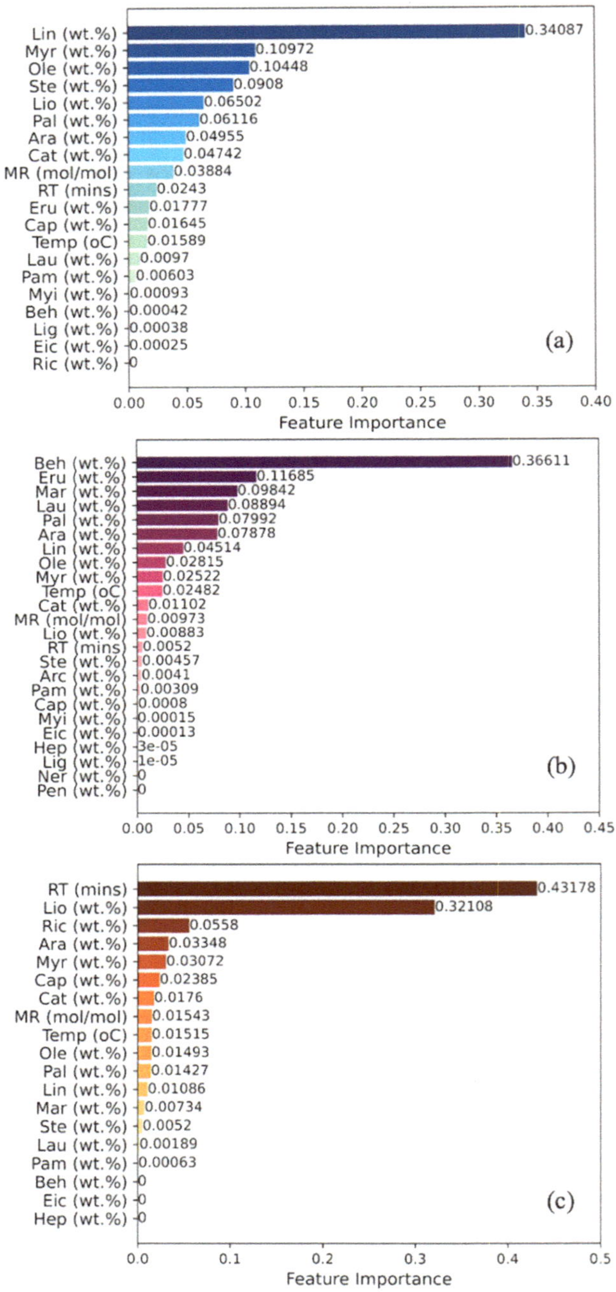

Fig. 6.4 1D-PDPs for operating condition features in predicting %BDY for **a** case 1, **b** case 2, and **c** case 3, from Sukpancharoen et al. (2023) (with permission)

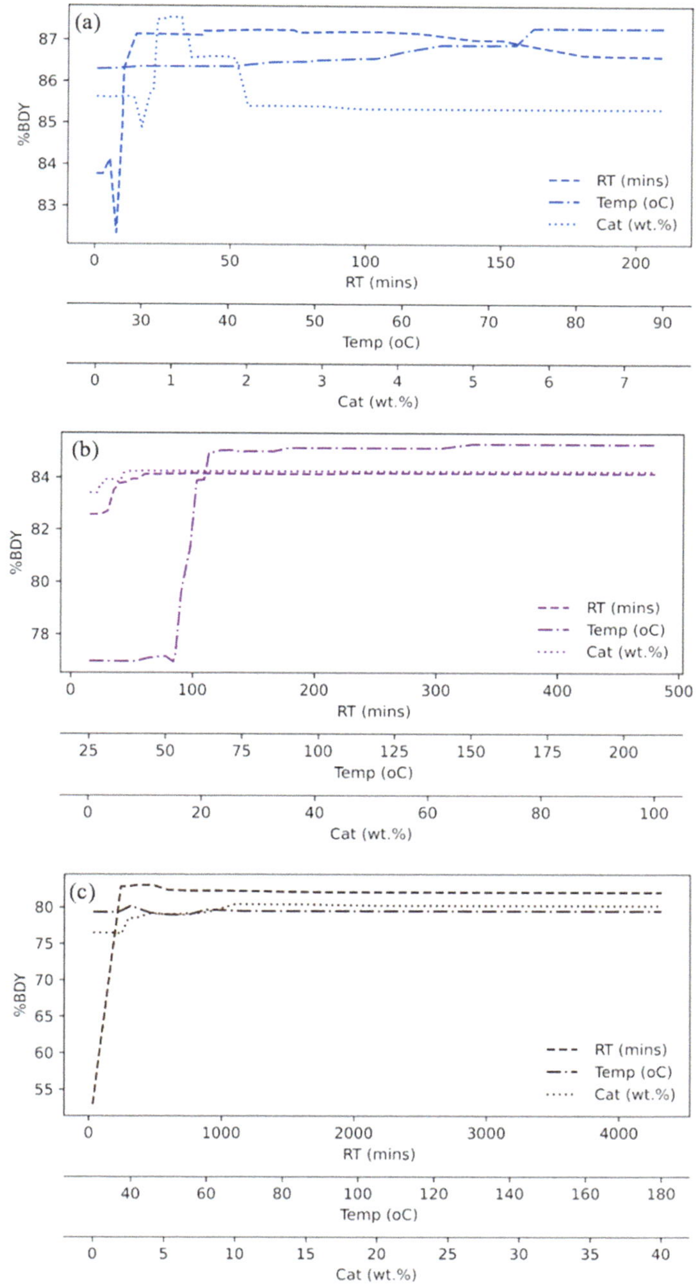

Fig. 6.5 Results of the 2D-PDPs for %BDY under two different operating conditions featuring **a** case 1, **b** case 2, and **c** case 3, from Sukpancharoen et al. (2023) (with permission)

time), while Fig. 6.5b, d, and f depict the relationship between MR and Temp (temperature) for homogeneous, heterogeneous, and enzymatic catalysts, respectively.

For both homogeneous and heterogeneous catalysts, Fig. 6.5 reveals that increasing reaction time leads to higher %BDY, while insufficient reaction time results in incomplete transesterification and reduced yields. Transesterification, which converts triglycerides into fatty acid methyl esters (FAMEs) and glycerol using methanol, typically requires at least six moles of methanol per mole of oil for optimal yield. However, excessive alcohol can decrease %BDY by reducing catalyst contact with the oil due to dilution effects.

Enzyme-based catalysts demonstrate a unique interaction between MR and RT, as shown in Fig. 6.5e. Using a 3:1 MR instead of the conventional 6:1 ratio minimizes enzyme denaturation caused by alcohol, ensuring better catalytic performance. Enzymes immobilized within a matrix can tolerate MRs greater than 4:1 while maintaining yields comparable to those achieved with a 3:1 ratio. Immobilization protects the enzyme from direct alcohol contact, enabling functionality at higher alcohol concentrations and temperatures up to 60 °C. For non-immobilized enzymes, maintaining temperatures between 30 °C and 40 °C is optimal, as shown in Fig. 6.5f.

The interaction between MR and temperature for homogeneous and heterogeneous catalysts, as depicted in Fig. 6.5b and d, shows that increasing temperatures within the 60–70 °C range enhances %BDY by accelerating the reaction. However, higher temperatures also increase saponification risks as highlighted by Encinar et al. (2007). Elevated temperatures can lead to alcohol evaporation and pressure buildup in closed reactors, posing safety risks.

These findings highlight the need to optimize operational conditions—reaction time, molar ratio, and temperature—based on the catalyst type to maximize biodiesel yield while minimizing inefficiencies and safety concerns.

6.5 Conclusions

This chapter investigated how different catalysts (homogeneous, heterogeneous, and enzyme-based) and operating conditions affect the transesterification of oils to produce biodiesel, focusing on predicting biodiesel yield (%BDY). Four machine learning (ML) algorithms were used: Support Vector Regression (SVR), Random Forest (RF), Extreme Gradient Boosting (XGB), and Kernel Ridge Regression (KRR). Among these, the XGB model showed the best performance, accurately predicting biodiesel yield with high R2 values and low error rates.

The analysis found that the fatty acid composition of the oils was the most important factor affecting %BDY, more so than operating conditions like reaction time, temperature, and catalyst concentration. This highlights the importance of selecting the right oils when making biodiesel, in addition to optimizing the transesterification process.

This chapter demonstrates the potential of machine learning algorithms to improve biodiesel yield prediction. Future research could incorporate more advanced artificial intelligence (AI) techniques to achieve even better prediction accuracy and reliability in modeling complex chemical processes, ultimately leading to more efficient and sustainable biodiesel production.

References

Agarwal AK (2007) Biofuels (alcohols and biodiesel) applications as fuels for internal combustion engines. Progress Energy Combust Sci 33(3):233–271. ISSN 0360-1285. https://doi.org/10.1016/j.pecs.2006.08.003. https://www.sciencedirect.com/science/article/pii/S0360128506000384

Almohana A, Almojil S, Amin M, Alali A, Kamal M, Elkhatib S, Felemban B, Algarni M (2022) Theoretical investigation on optimization of biodiesel production using waste cooking oil: Machine learning modeling and experimental validation. Energy Rep 8:11938–11951, 11 2022. https://doi.org/10.1016/j.egyr.2022.08.265

Bounoua Z, Ouazzani Chahidi L, Mechaqrane A (2021) Estimation of daily global solar radiation using empirical and machine-learning methods: A case study of five moroccan locations. Sustain Mater Technol 28:e00261. ISSN 2214-9937. https://doi.org/10.1016/j.susmat.2021.e00261. https://www.sciencedirect.com/science/article/pii/S2214993721000166

Chen T, Guestrin C (2016) Xgboost: A scalable tree boosting system. In: Proceedings of the 22nd ACM SIGKDD international conference on knowledge discovery and data mining, KDD 16, pp 785 – 794, New York, NY, USA. Association for Computing Machinery. ISBN 9781450342322. https://doi.org/10.1145/2939672.2939785

Dutta MK, Kaur M, Sarkar RK (2022) Image quality improvement of old and distorted artworks using fuzzy logic technique. Optik 249:168252. ISSN 0030-4026. https://doi.org/10.1016/j.ijleo.2021.168252. https://www.sciencedirect.com/science/article/pii/S0030402621017757

Elaskary RM, Mehana AH, Fahmy Y, El-Ghoneimy M (2022) Closed-form mse and achievable rate for indirect learning dpd. Phys Commun 54:101801. ISSN 1874-4907. https://doi.org/10.1016/j.phycom.2022.101801. https://www.sciencedirect.com/science/article/pii/S1874490722001148

Encinar J, Gonzalez J, Rodriguez-Reinares A (2007) Ethanolysis of used frying oil. Biodiesel preparation and characterization. Fuel Proc Technol 88(5):513–522. ISSN 0378-3820. https://doi.org/10.1016/j.fuproc.2007.01.002. https://www.sciencedirect.com/science/article/pii/S0378382007000045

Endalew AK, Kiros Y, Zanzi R (2011) Inorganic heterogeneous catalysts for biodiesel production from vegetable oils. Biomass Bioenergy 35(9):3787–3809. ISSN 0961-9534. https://doi.org/10.1016/j.biombioe.2011.06.011. https://www.sciencedirect.com/science/article/pii/S0961953411003357

Entezari A, Aslani A, Zahedi R, Noorollahi Y (2023) Artificial intelligence and machine learning in energy systems: A bibliographic perspective. Energy Strat Rev 45:101017. ISSN 2211-467X. https://doi.org/10.1016/j.esr.2022.101017. https://www.sciencedirect.com/science/article/pii/S2211467X22002115

Fang Y, Ma L, Yao Z, Li W, You S (2022) Process optimization of biomass gasification with a monte carlo approach and random forest algorithm. Energy Conv Manag 264:115734. ISSN 0196-8904. https://doi.org/10.1016/j.enconman.2022.115734. https://www.sciencedirect.com/science/article/pii/S0196890422005301

Fentis A, Rafik M, Bahatti L, Bouattane O, Mestari M (2022) Data driven approach to forecast the next day aggregate production of scattered small rooftop solar photovoltaic systems without meteorological parameters. Energy Rep 8:3221–3233. ISSN 2352-4847. https://doi.org/10.1016/j.egyr.2022.02.088. https://www.sciencedirect.com/science/article/pii/S2352484722003353

Gupta KK, Kalita K, Ghadai RK, Ramachandran M, Gao X-Z (2021) Machine learning-based predictive modelling of biodiesel production - a comparative perspective. Energies 14(4). ISSN 1996-1073. https://doi.org/10.3390/en14041122. https://www.mdpi.com/1996-1073/14/4/1122

Hai A, Bharath G, Patah MFA, Daud WMAW, K R, Show P, Banat F (2023) Machine learning models for the prediction of total yield and specific surface area of biochar derived from agricultural biomass by pyrolysis. Environ Technol I& Innov 30:103071. ISSN 2352-1864. https://doi.org/10.1016/j.eti.2023.103071. https://www.sciencedirect.com/science/article/pii/S2352186423000676

Halay N, Todros K (2019) Mse based optimization of the measure-transformed music algorithm. Signal Proc 160:150–163. ISSN 0165-1684. https://doi.org/10.1016/j.sigpro.2019.01.025. https://www.sciencedirect.com/science/article/pii/S0165168419300490

Hoang AT (2021) Prediction of the density and viscosity of biodiesel and the influence of biodiesel properties on a diesel engine fuel supply system. J Marine Eng I& Technol 20(5):299–311. https://doi.org/10.1080/20464177.2018.1532734

Jayakumar M, Karmegam N, Gundupalli MP, Bizuneh Gebeyehu K, Tessema Asfaw B, Chang SW, Ravindran B, Kumar Awasthi M (2021) Heterogeneous base catalysts: synthesis and application for biodiesel production - a review. Bioresour Technol 331:125054. ISSN 0960-8524. https://doi.org/10.1016/j.biortech.2021.125054. https://www.sciencedirect.com/science/article/pii/S096085242100393X

Jentzer JC, Kashou AH, Murphree DH (2023) Clinical applications of artificial intelligence and machine learning in the modern cardiac intensive care unit. Intell-Based Med 7:100089. ISSN 2666-5212. https://doi.org/10.1016/j.ibmed.2023.100089. https://www.sciencedirect.com/science/article/pii/S2666521223000030

Jeong G-T, Park D-H (2009) Optimization of biodiesel production from castor oil using response surface methodology. Appl Biochem Biotechnol 156(1):1–11. ISSN 1559-0291. https://doi.org/10.1007/s12010-008-8468-9

Kang H, Zong X, Wang J, Chen H (2023) Binary gravity search algorithm and support vector machine for forecasting and trading stock indices. Int Rev Econ I& Finance 84:507–526. ISSN 1059-0560. https://doi.org/10.1016/j.iref.2022.11.009. https://www.sciencedirect.com/science/article/pii/S1059056022002775

Karunasingha DSK (2022) Root mean square error or mean absolute error? use their ratio as well. Inf Sci 585:609–629. ISSN 0020-0255. https://doi.org/10.1016/j.ins.2021.11.036. https://www.sciencedirect.com/science/article/pii/S0020025521011567

Katongtung T, Onsree T, Tippayawong N (2022) Machine learning prediction of biocrude yields and higher heating values from hydrothermal liquefaction of wet biomass and wastes. Bioresour Technol 344:126278. ISSN 0960-8524. https://doi.org/10.1016/j.biortech.2021.126278. https://www.sciencedirect.com/science/article/pii/S0960852421016205

Krishnamurthy K, Sridhara S, Ananda Kumar C (2018) Synthesis and optimization of hydnocarpus wightiana and dairy waste scum as feed stock for biodiesel production by using response surface methodology. Energy 153:1073–1086. ISSN 0360-5442. https://doi.org/10.1016/j.energy.2018.04.068. https://www.sciencedirect.com/science/article/pii/S0360544218306753

Kumar D, Das T, Giri BS, Verma B (2020) Preparation and characterization of novel hybrid bio-support material immobilized from pseudomonas cepacia lipase and its application to enhance biodiesel production. Renew Energy 147:11–24. ISSN 0960-1481. https://doi.org/10.1016/j.renene.2019.08.110. https://www.sciencedirect.com/science/article/pii/S0960148119312972

Le H-A, Le D-A, Le T-T, Le H-P, Le T-H, Hoang H-GT, Nguyen T-A (2022) An extreme gradient boosting approach to estimate the shear strength of frp reinforced concrete beams. Structures 45:1307–1321. ISSN 2352-0124. https://doi.org/10.1016/j.istruc.2022.09.112. https://www.sciencedirect.com/science/article/pii/S2352012422008888

Ma F, Hanna MA (1999) Biodiesel production: a review. Bioresour Technol 70(1):1–15. ISSN 0960-8524. https://doi.org/10.1016/S0960-8524(99)00025-5. https://www.sciencedirect.com/science/article/pii/S0960852499000255

Mathiyazhagan M, Ganapathi A (2011) Factors affecting biodiesel production. Res Plant Biol 1(2):1–5

Moayedi H, Aghel B, Foong LK, Bui DT (2020) Feature validity during machine learning paradigms for predicting biodiesel purity. Fuel 262:116498. ISSN 0016-2361. https://doi.org/10.1016/j.fuel.2019.116498. https://www.sciencedirect.com/science/article/pii/S0016236119318526

Modiba E, Osifo P, Rutto H (2014) Biodiesel production from baobab (adansonia digitata l.) seed kernel oil and its fuel properties. Ind Crops Prod 59:50–54. ISSN 0926-6690. https://doi.org/10.1016/j.indcrop.2014.04.044. https://www.sciencedirect.com/science/article/pii/S0926669014002593

Onsree T, Tippayawong N (2021) Machine learning application to predict yields of solid products from biomass torrefaction. Renew Energy 167:425–432. ISSN 0960-1481. https://doi.org/10.1016/j.renene.2020.11.099. https://www.sciencedirect.com/science/article/pii/S0960148120318516

Panapitiya G, Avendano-Franco G, Ren P, Wen X, Li Y, Lewis JP (2018) Machine-learning prediction of co adsorption in thiolated, ag-alloyed au nanoclusters. J Am Chem Soc 140(50):17508–17514

Phromphithak S, Onsree T, Tippayawong N (2021) Machine learning prediction of cellulose-rich materials from biomass pretreatment with ionic liquid solvents. Bioresour Technol 323:124642. ISSN 0960-8524. https://doi.org/10.1016/j.biortech.2020.124642. https://www.sciencedirect.com/science/article/pii/S0960852420319167

Pinzi S, Lopez-Gimenez F, Ruiz J, Dorado M (2010) Response surface modeling to predict biodiesel yield in a multi-feedstock biodiesel production plant. Bioresour Technol 101(24):9587–9593. ISSN 0960-8524. https://doi.org/10.1016/j.biortech.2010.07.076. https://www.sciencedirect.com/science/article/pii/S0960852410012666

Qu Z, Xu J, Wang Z, Chi R, Liu H (2021) Prediction of electricity generation from a combined cycle power plant based on a stacking ensemble and its hyperparameter optimization with a grid-search method. Energy 227:120309. ISSN 0360-5442. https://doi.org/10.1016/j.energy.2021.120309. https://www.sciencedirect.com/science/article/pii/S0360544221005582

Rashid U, Anwar F, Ashraf M, Saleem M, Yusup S (2011) Application of response surface methodology for optimizing transesterification of moringa oleifera oil: Biodiesel production. Energy Conver Manag 52(8):3034–3042. ISSN 0196-8904. https://doi.org/10.1016/j.enconman.2011.04.018. https://www.sciencedirect.com/science/article/pii/S0196890411001488

Sanjika Devi R, Bindu K, Kurup DG (2022) Behavioral modeling and digital predistortion of rf power amplifiers based on time-delay kernel ridge regression. AEU - Int J Electr Commun 152:154239. ISSN 1434-8411. https://doi.org/10.1016/j.aeue.2022.154239. https://www.sciencedirect.com/science/article/pii/S1434841122001364

Sharma P, Sahoo BB, Said Z, Hadiyanto H, Nguyen XP, Nizetic S, Huang Z, Hoang AT, Li C (2023) Application of machine learning and box-behnken design in optimizing engine characteristics operated with a dual-fuel mode of algal biodiesel and waste-derived biogas. Int J Hydr Energy 48(18):6738–6760. ISSN 0360-3199. https://doi.org/10.1016/j.ijhydene.2022.04.152. https://www.sciencedirect.com/science/article/pii/S0360319922017244. The 5th International Conference on Alternative Fuel, Energy I& Environment (ICAFEE 2021)

Shi Y, Zhao G, Wang M, Xu Y (2022) An adaptive grid search algorithm for fitting spherical target of terrestrial lidar. Measurement 198:111430. ISSN 0263-2241. https://doi.org/10.1016/j.measurement.2022.111430. https://www.sciencedirect.com/science/article/pii/S0263224122006601

Sukpancharoen S, Katongtung T, Rattanachoung N, Tippayawong N (2023) Unlocking the potential of transesterification catalysts for biodiesel production through machine learning approach 378:128961. https://doi.org/10.1016/j.biortech.2023.128961

Tacias-Pascacio VG, Torrestiana-Sanchez B, Dal Magro L, Virgen-Ortiz JJ, Suarez-Ruiz FJ, Rodrigues RC, Fernandez-Lafuente R (2019) Comparison of acid, basic and enzymatic catalysis on the production of biodiesel after rsm optimization. Renew Energy 135:1–9. ISSN 0960-1481. https://doi.org/10.1016/j.renene.2018.11.107. https://www.sciencedirect.com/science/article/pii/S0960148118314265

Tuan Hoang A, Nizetic S, Chyuan Ong H, Tarelko W, Viet Pham V, Hieu Le T, Quang Chau M, Phuong Nguyen X (2021) A review on application of artificial neural network (ann) for performance and emission characteristics of diesel engine fueled with biodiesel-based fuels. Sustain Energy Technol Assess 47:101416. ISSN 2213-1388. https://doi.org/10.1016/j.seta.2021.101416. https://www.sciencedirect.com/science/article/pii/S2213138821004264

Vasic K, Hojnik Podrepsek G, Knez Z, Leitgeb M (2020) Biodiesel production using solid acid catalysts based on metal oxides. Catalysts 10(2). ISSN 2073-4344. https://doi.org/10.3390/catal10020237. https://www.mdpi.com/2073-4344/10/2/237

Vu HL, Ng KTW, Richter A, An C (2022) Analysis of input set characteristics and variances on k-fold cross validation for a recurrent neural network model on waste disposal rate estimation. J Environ Manag 311:114869. ISSN 0301-4797. https://doi.org/10.1016/j.jenvman.2022.114869. https://www.sciencedirect.com/science/article/pii/S030147972200442X

Yucel Y (2012) Optimization of biocatalytic biodiesel production from pomace oil using response surface methodology. Fuel Proc Technol 99:97–102. ISSN 0378-3820. https://doi.org/10.1016/j.fuproc.2012.02.008. https://www.sciencedirect.com/science/article/pii/S0378382012000823

Zabeti M, Wan Daud WMA, Aroua MK (2009) Activity of solid catalysts for biodiesel production: A review. Fuel Proc Technol 90(6):770–777. ISSN 0378-3820. https://doi.org/10.1016/j.fuproc.2009.03.010. https://www.sciencedirect.com/science/article/pii/S0378382009000629

Zhang X, Liu C-A (2023) Model averaging prediction by k-fold cross-validation. J Econom 235(1):280–301. ISSN 0304-4076. https://doi.org/10.1016/j.jeconom.2022.04.007. https://www.sciencedirect.com/science/article/pii/S0304407622000975

Zhao R, Yang X, Li M, Peng X, Wei M, Zhang X, Yang L, Li J (2021) Biodiesel preparation from thlaspi arvense l. Seed oil utilizing a novel ionic liquid core-shell magnetic catalyst. Ind Crops Prod 162:113316. ISSN 0926-6690. https://doi.org/10.1016/j.indcrop.2021.113316. https://www.sciencedirect.com/science/article/pii/S0926669021000807

Zhao X, Yang H, Yao Y, Qi H, Guo M, Su Y (2022) Factors affecting traffic risks on bridge sections of freeways based on partial dependence plots. Phys A: Stat Mech Appl 598:127343. ISSN 0378-4371. https://doi.org/10.1016/j.physa.2022.127343. https://www.sciencedirect.com/science/article/pii/S0378437122002709

Zhu L, Zhou X, Zhang C (2021) Rapid identification of high-quality marine shale gas reservoirs based on the oversampling method and random forest algorithm. Artif Intell Geosci 2:76–81. ISSN 2666-5441. https://doi.org/10.1016/j.aiig.2021.12.001. https://www.sciencedirect.com/science/article/pii/S2666544121000307

Chapter 7
Anaerobic Digestion

Abstract Machine learning (ML) techniques can be used to predict the specific methane yields (SMY) of lignocellulosic biomass (LB) by analyzing a 14-feature dataset. These features cover both the characteristics of the biomass and the operating conditions of continuously fed, completely mixed reactors. Among the ML models tested, the random forest (RF) model performed the best, achieving a high level of predictive accuracy with a coefficient of determination (R^2) of 0.85 and a root mean square error (RMSE) of 0.06. The composition of the biomass was found to have a significant impact on methane yields. Cellulose emerged as the most important factor, outweighing lignin and the biomass ratio in terms of influence. Using the RF model, the effect of the ratio of lignocellulosic biomass to manure was examined to determine the optimal conditions for biogas production. It was found that, under standard organic loading rates (OLR), a 1:1 ratio of lignocellulosic biomass to manure yielded the best results. Experimental tests confirmed these findings, with the highest SMY reaching 79.2% of the value predicted by the RF model. This chapter highlights how machine learning can be effectively applied to model and optimize anaerobic digestion processes, particularly when working with lignocellulosic biomass. By leveraging ML techniques, valuable insights are gained into the factors that influence methane production and improve the efficiency of biogas generation.

Keywords Machine learning · Co-digestion · Specific methane yield · Lignocellulosic biomass · Feedstock ratio

7.1 Introduction

Energy demands have risen significantly due to population growth and industrial expansion (Ram et al. 2019). Renewable energy sources offer a promising solution, with energy crops and agricultural residues showing great potential as feedstocks for biogas production due to their widespread availability (Potter et al. 2012). Among these, lignocellulosic biomass (LB) is particularly advantageous because it can be

N. Tippayawong et al., *Machine Learning Applications for Bioenergy Conversion*, Green Energy and Technology, https://doi.org/10.1007/978-981-96-8086-3_7

managed to ensure a consistent supply and quality, and it is easily accessible in agricultural regions (Kasinath et al. 2021). Lignocellulosic biomass refers to plant-based material composed of cellulose, hemicellulose, and lignin. It is the most abundant raw material on Earth for biofuel production. Lignocellulosic biomass is categorized into virgin biomass, waste biomass, and energy crops. Virgin biomass includes plants, while waste biomass is a byproduct from agriculture (like corn stover and sugarcane bagasse) and forestry (sawmill and paper mill discards). Energy crops, such as switchgrass, are specifically grown for lignocellulosic biomass production to generate sustainable energy.

Research has explored biogas production from various types of LB, such as rice straw, corn stover, giant juncao grass, and perennial ryegrass, using continuously stirred-tank reactors. These studies have reported specific methane yields (SMYs) ranging from 84 to 405 mL CH_4 g VS^{-1} (Li et al. 2014, 2015; Sonwai et al. 2021; Voelklein et al. 2016). Volatile solids (VS) represent the portion of the total solids that are organic matter, excluding inorganic salts and ash. Since bacterial action in anaerobic digestion primarily occurs on organic matter, the VS content indicates the potential for biogas production from a particular feedstock. Generally, materials high in volatile solids, such as fats, produce more biogas compared to materials like manure, which have lower volatile solids content.

While LB holds significant potential for biogas production, variations in its properties and composition can lead to differences in biogas yields. Biogas production can be carried out through mono-digestion or co-digestion. Mono-digestion often faces challenges, such as nutrient deficiencies, which can limit microbial activity over time (FitzGerald et al. 2019). In contrast, co-digestion can enhance the degradation process by combining substrates to create synergistic effects (Karki et al. 2021). For optimal biogas production, the recommended biomass-to-manure ratio ranges from 1:1 to 4:1 (on a volatile solids basis), yielding SMYs of 223–398 mL CH_4 g VS^{-1} at an organic loading rate (OLR) of 3.5–4.0 kg VSm^{-3} d^{-1}. However, many factors, including temperature, pH, volatile fatty acids (VFA), and substrate proportions influence biogas production. Therefore, finding the ideal balance for each substrate combination is crucial.

Machine learning (ML), a branch of artificial intelligence (AI), has proven to be a powerful tool for predicting and optimizing biogas production. ML models analyze datasets to identify patterns and make predictions. Compared to traditional modeling approaches, ML is particularly effective at handling the complex, nonlinear behaviors typical of anaerobic digestion (AD) processes (Khan et al. 2023; Wang et al. 2023). Recent reviews have highlighted the use of ML in AD for tasks such as process optimization, parameter prediction, and disturbance detection (Andrade Cruz et al. 2022; Gupta et al. 2023). Additionally, ML can accelerate research by reducing the time, cost, and resources required for experimentation (Meena et al. 2021). This chapter explores how ML can be applied to improve our understanding and optimization of biogas production from lignocellulosic biomass.

Popular machine learning (ML) algorithms for anaerobic digestion (AD) include artificial neural networks (ANN), support vector machines (SVM), random forests (RF), and extreme gradient boosting (XGBoost). These models are commonly used to predict variables such as biogas yield, process stability, and effluent characteristics. For example, ANN achieved a high coefficient of determination (R^2) of 0.98 when predicting specific methane yield (SMY) for the co-digestion of agricultural wastes and cattle manure (Almomani 2020). However, its performance was less effective (R^2 = 0.64) for the co-digestion of lignocellulosic biomass (LB) with cattle manure (Li et al. 2016). On the other hand, RF and XGBoost have demonstrated strong predictive performance, with R^2 values of 0.80 and 0.88, respectively, for the co-digestion of mixed agricultural, industrial, and municipal wastes (Djavan et al. 2020). While ANN is widely used, its "black-box" nature, lack of interpretability, and need for large datasets are notable limitations (Khan et al. 2023). In contrast, tree-based models like RF and XGBoost are often preferred because they are easier to use and provide clear insights into feature importance. Additionally, kernel ridge regression (KRR) has proven effective for small datasets, outperforming ANN, SVM, and decision trees in modeling SMY for the mono-digestion of LB (Wang et al. 2023).

Despite these advancements, comprehensive ML applications for analyzing AD performance specifically for LB are still limited. Many studies focus on either feedstock characteristics or operational conditions but rarely consider both. Recent research on LB anaerobic digestion has often overlooked co-digestion or thermophilic biodegradation, and the robustness of the models has not always been validated through experiments (Khan et al. 2023). Furthermore, the potential of ML to determine the optimal biomass-to-manure ratio for co-digestion remains underexplored. Given the complexity of biogas production from LB, understanding the interplay of relevant parameters is crucial for optimizing the process.

This chapter aims to achieve the following objectives:

1. Develop models using RF, XGBoost, and KRR to predict SMY from LB based on 14 input features, including LB characteristics and operating conditions.
2. Identify the key LB compositions, operational conditions, and interactions that influence AD performance.
3. Determine the optimal biomass-to-manure ratio for co-digestion at various organic loading rates (OLRs).
4. Validate ML predictions through pilot-scale AD experiments using fresh giant juncao grass, dried rice straw, and dried corn stover as feedstocks in both mono-digestion and co-digestion with cattle manure.

By addressing these goals, this chapter seeks to advance the application of ML in optimizing biogas production from lignocellulosic biomass, providing valuable insights for both researchers and practitioners in the field.

7.2 Materials and Methods

A total of 275 datasets related to biogas production from lignocellulosic biomass (LB) were collected from previous studies. Only experiments conducted under continuous or semi-continuous modes of operation were included to ensure that the input features

Table 7.1 Set of input features and target variable, from Sonwai et al. (2023) (with permission)

Variable	Units	Name	Values	Average	Standard deviation	Constraints[a]
Input features/operating conditions						
Ratio of biomass[b]	%	Bio	5.00–100	69.3	34.5	x
Organic loading rates	kg VS $m^{-3}d^{-1}$	OLR	0.50–7.00	2.60	1.30	1.0–4.0
Biomass size	mm	Size	0.30–60.0	14.5	12.5	10.0
Volume of reactor	L	Vol	1.10–50.0	9.40	13.2	50.0
Hydraulic retention time	d	HRT	2.00–92.0	31.1	20.0	50.0
Temperature	°C	T	20.0–55.0	37.8	6.50	55.0
Volatile solid	%	VS	53.5–97.6	79.1	12.6	80.0
pH	–	pH	3.80–8.20	7.30	0.60	7.00
Biomass characteristics						
Carbon content	%	C	3.30–81.5	44.5	12.2	46.3
Nitrogen content	%	N	0.50–10.2	1.70	1.0	0.70
Carbon to Nitrogen ratio	–	CN	3.60–74.8	33.0	17.5	69.9
Cellulose	%	Cel	9.00–45.5	31.8	10.9	41.2
Hemicellulose	%	Hemi	9.80–32.6	22.4	5.50	21.5
Lignin	%	Lig	2.40–42.0	9.30	7.90	18.2
Target variable						
Specific methane yield	$mL\,CH_4 g\,VS^{-1}$	SMY	39.0–414	226.9	80.8	y

Note [a]Constraints are parameters from previous studies (Sonwai et al. 2021) that are set for use in the optimization process. [b]Ratio of biomass is the proportion of a biomass (VS basis) used in co-digestion with manure

aligned with typical anaerobic digestion (AD) processes. After removing datasets with incomplete values, 205 complete datasets from 30 publications were retained for analysis.

The dataset included 14 input features (Table 7.1), which were divided into two categories:

- Operating conditions and biomass characteristics. The operating conditions included biomass ratio (Bio), organic loading rate (OLR), biomass size (Size), reactor volume (Vol), hydraulic retention time (HRT), temperature (T), volatile solids (VS), and pH.
- Biomass characteristics included carbon content (C), nitrogen content (N), carbon-to-nitrogen ratio (CN), cellulose (Cel), hemicellulose (Hemi), and lignin (Lig).

The target variable was the specific methane yield (SMY). To better understand the data, box plots and histograms were created to visualize the distributions of the input features and SMY values. Previous experimental data from Sonwai et al. (2021) were used as a reference to determine the optimal co-digestion ratio between biomass and manure. To prepare the dataset for analysis, Yeo-Johnson's nonlinear transformation (In-Kwon and A. 2000) was applied for standardization and normalization, as detailed in Sect. 2.7. The Pearson correlation coefficient was used to measure the strength of linear relationships between variables, as described in Sect. 2.7.1. This approach ensures that the data is properly prepared and analyzed, providing a solid foundation for the machine learning models used in this study.

7.2.1 *Machine Learning Algorithms Selection and Tuning*

The random forest (RF), extreme gradient boosting (XGBoost), and kernel ridge regression (KRR) algorithms were selected to determine which one best suited the dataset. These algorithms were implemented using the scikit-learn library for Python (Pedregosa et al. 2011). The dataset was split into training and testing groups in an 80:20 ratio.

Hyperparameter optimization for each algorithm was carried out using a full-factor grid search combined with tenfold cross-validation. Once the best hyperparameters were identified, the models were retrained and tested on the remaining 20% of the data. The performance of each model was evaluated using two metrics: the coefficient of determination (R^2) and the root mean square error (RMSE) (Phromphithak et al. 2021).

R^2 measures how well the independent variables explain the variance in the dependent variable. An R^2 value closer to 1 indicates higher model accuracy. RMSE, on the other hand, is the square root of the average squared prediction errors and measures the accuracy of predictions. Lower RMSE values indicate better performance, with values close to zero being ideal. These metrics provide a comprehensive assessment of the models' predictive capabilities, helping to identify the most effective algorithm for the dataset.

7.2.2 *Feature Evaluation*

To understand how each input feature influenced the specific methane yield (SMY) predictions, the 14 input features were analyzed in two scenarios:

- In Case 1, all 14 input features were included in the analysis.
- In Case 2, cellulose (Cel), hemicellulose (Hemi), and lignin (Lig) were excluded to assess the impact of these critical biomass components on SMY predictions.

For both cases, hyperparameters were optimized using a full-factor grid search combined with tenfold cross-validation. By prioritizing the input features, machine learning (ML) algorithms identified the most important variables affecting SMY predictions. These prioritized features provided valuable insights into the dynamics of biogas production, helping to optimize anaerobic digestion (AD) processes for lignocellulosic biomass (LB).

Partial dependence plots (PDPs) were used to explore the relationships between input features and SMY predictions. One-way PDPs illustrated how individual high-priority features influenced the target variable, while two-way PDPs highlighted interactions between pairs of input features. These two-way PDPs helped identify conditions where input features interacted to optimize biogas production, offering a deeper understanding of process dynamics and potential improvements for LB-based AD systems. This approach not only clarifies the role of individual features but also reveals how their interactions can be leveraged to enhance biogas yields.

7.2.3 *Optimization of the Proportion of Co-digestion Process*

Enhancing the stability of the anaerobic digestion (AD) process for lignocellulosic biomass (LB) can be achieved by co-digesting it with manure. The random forest (RF) model, developed from the most accurate dataset, was chosen to optimize the co-digestion process. In this approach, the biomass ratio (Bio) was defined as the independent variable (x), and specific methane yield (SMY) was set as the target variable (y). The remaining 13 parameters were treated as predetermined control variables or constraints, based on optimal conditions reported in previous studies (Table 7.1).

The optimization process used the RF model, running each feature independently across various scenarios through machine learning operations in Python with the scikit-learn library. To ensure practical application in reactor operations, the model was tested with varying organic loading rates (OLR): 1.0, 1.5, 2.0, 2.5, 3.0, 3.5, and 4.0 kg VS $m^{-3}d^{-1}$. For each OLR, the optimal biomass-to-manure ratio was determined. These results provided a foundation for designing AD experiments, ensuring that the findings could be effectively compared and applied in real-world reactor operations. This approach not only enhances the stability of the AD process but also offers practical insights for optimizing biogas production from lignocellulosic biomass.

7.3 Anaerobic Digestion Experiments

Experiments were conducted in lab-scale biogas production. They used three identical continuously stirred-tank reactors (CSTRs) made of stainless steel, each with an effective volume of 50 liters. The reactors were equipped with heating rods to maintain mesophilic (35 °C ± 1 °C) or thermophilic (55 °C ± 1 °C) conditions. Mixing was achieved using a paddle system operated at 150 rpm for 5 min, followed by a 15-minute pause. The reactors were fed with ground Giant Juncao Grass (GJG), which was shredded to a particle size of less than 10 mm and stored at −20 °C before use. Effluent recirculation was implemented to manage solid content, and biogas was collected in PVC bags for analysis. The experiment also included microbial community analysis to understand the relationship between operating conditions and microbial activity, providing a comprehensive evaluation of the factors influencing biogas production from GJG. This aims to bridge the gap between lab-scale experiments and full-scale applications, offering practical insights for optimizing biogas production from lignocellulosic biomass.

7.3.1 Inoculum and Biomass

The inoculum used for the experiments was collected from a full-scale biogas reactor that processes cattle manure (CM) at the Chiang Mai Fresh Milk Farm in Chiang Mai, Thailand. Three types of biomass commonly grown in the region—Giant Juncao Grass (GJG), rice straw (RS), and corn stover (CS)—were chosen as feedstocks for the study.

The biomass was ground into small particles, each less than 10 mm in size, using a heavy-duty mincer (CLEO blender model CCB403/404). The cattle manure (CM) was sourced from an agricultural area in Sansai District, Chiang Mai, Thailand. To preserve their quality before use, GJG and CM were stored at −20 °C and 4 °C, respectively, while RS and CS were kept at room temperature (25–27 °C). The key characteristics of the feedstocks and the inoculum are summarized in Table 7.2.

7.3.2 Experimental Conditions and Set-Up

The experiments were carried out using three 50-L continuously stirred tank reactors (CSTRs), designated as the GJG reactor (GR), RS reactor (RR), and CS reactor (CR). These reactors operated under thermophilic conditions (55.4 ± 0.4 °C). For microbial

Table 7.2 Characteristics of feedstocks and inoculum, from Sonwai et al. (2023) (with permission)

Parameters	Units	GJG	RS	CS	CM	Inoculum
Density	$kg\,m^{-3}$	890.2 ± 16.3	275 ± 73.0	574 ± 194.1	846 ± 11.0	–
Moisture content	%	79.3 ± 0.58	5.72 ± 1.93	7.22 ± 3.16	74.0 ± 2.26	–
SS	$mg\,L^{-1}$	–	–	–	–	21,013 ± 1,642
VSS	$mg\,L^{-1}$	–	–	–	–	16,387 ± 1,838
TS	%	21.2 ± 0.77	94.7 ± 1.47	93.5 ± 2.46	25.5 ± 3.56	2.71 ± 0.04
VS	%	18.8 ± 0.84	81.7 ± 1.47	81.9 ± 1.87	16.5 ± 0.25	1.94 ± 0.02
Organic Matter	%	58.6 ± 1.61	57.9 ± 1.04	62.4 ± 1.33	54.5 ± 1.37	–
Nitrogen	%	1.04 ± 0.07	1.32 ± 0.04	1.77 ± 0.03	2.33 ± 0.24	–
C/N ratio	–	32.7 ± 0.69	25.5 ± 0.55	20.4 ± 1.10	13.6 ± 0.68	–
Cellulose	%	46.6 ± 1.80	44.1 ± 1.00	37.3 ± 0.60	21.8 ± 1.13	–
Hemicellulose	%	26.4 ± 1.07	25.8 ± 1.72	24.4 ± 0.56	9.42 ± 0.90	–
Lignin	%	17.8 ± 0.97	9.78 ± 0.24	16.6 ± 1.41	53.5 ± 1.36	–

Note n = 3; Results are shown as Mean ± SD

Table 7.3 Details of all experimental conditions, from Sonwai et al. (2023) (with permission)

Condition	Reactors	Feedstock ratio (Biomass: CM)	OLR ($kg\,VS\,m^{-3}d^{-1}$)	Trace element addition	Effluent recirculation rate (Effluent: Biomass by volume)
Mono-digestion	GR	1:0	1.0–4.0	Yes	3:1
	RR	1:0	1.0–4.0	Yes	3:1
	CR	1:0	1.0–4.0	Yes	3:1
Co-digestion	GR	1:1	4.0	No	3:1
	RR	1:1	4.0	No	3:1
	CR	1:1	4.0	No	3:1

acclimatization, all reactors were firstly fed with the ground fresh Napier Pak Chong grass for 150 days at an OLR of 1.0 $kg\,VS\,m^{-3}d^{-1}$ under thermophilic (55.0 ± 1 °C) temperatures. After attaining stable methane production, the ground fresh GJG was used as the feedstock for all studied reactors at OLR of 1.5 $kg\,VS\,m^{-3}d^{-1}$ for 25 days following by the OLR of 2.0 $kg\,VS\,m^{-3}d^{-1}$ for 20 days, respectively. During the start-up period, the mixing system was turned off for at least 30 min before taking off the effluent to ensure that the reactor had sufficient amounts of microbes for semi-continuous operation (Sonwai et al. 2021).

After the start-up phase, each reactor was fed its respective mono-substrate (GJG, RS, or CS) at organic loading rates (OLR) of 1.0, 2.0, 3.0, and 4.0 $kg\,VS\,m^{-3}d^{-1}$. To ensure sufficient micronutrients, trace elements were added to the feed at the following concentrations: 500 $mg\,L^{-1}$; $FeCl_3 - 6H_2O$; 5 $mg\,L^{-1}$ $NiCl_2 - 6H_2O$ and 0.5 $mg\,L^{-1}CoCl_2 - 6H_2O$. When the reactors reached steady state at OLR 4.0 $kg\,VS\,m^{-3}d^{-1}$, co-digestion was initiated using a 1:1 (VS basis) biomass-to-CM ratio, without the addition of trace elements. The reactors were fed twice daily, with an effluent recirculation rate of 3:1 (by volume of effluent to feedstock). This experimental setup was maintained for 329 days. A summary of the experimental conditions is provided in Table 7.3.

7.3.3 Analytical Methods

The influent refers to the feedstock mixture—in this case, Giant Juncao Grass (GJG)—that is introduced into the reactor. The characteristics of the influent, such as its volatile solids (VS) content, particle size, and chemical composition, directly impact the efficiency of the digestion process. Effluent, on the other hand, is the liquid and solid residue that remains after the digestion process. Analyzing the effluent provides valuable insights into the reactor's performance, including the extent of biomass degradation and the stability of the digestion process. By recirculating a portion of the effluent back into the reactor, the system maintains a balanced solid concentration, which is crucial for sustaining microbial activity and optimizing biogas production.

Both influent and effluent samples were regularly analyzed to monitor key parameters such as pH, volatile fatty acids (VFA), alkalinity, and nutrient levels. These analyses help ensure that the reactors operate under optimal conditions and provide data to assess the impact of variables like temperature, organic loading rate (OLR), and recirculation rates on biogas yield and process stability. Influent and effluent samples were collected regularly to assess the performance of the reactors. Effluent pH was measured daily using a pH meter (pH Testr-30, Eutech). Total solids (TS) and volatile solids (VS) in both influent and effluent were analyzed twice a week, following standard methods. Effluent alkalinity (Alk) and volatile fatty acids (VFA) were measured using the direct titration method as described by DiLallo and Albertson (1961).

Biogas was collected in bags, with its volume measured daily using an oil rotary vacuum pump (Type RP-S 100H, MAKASHI, Japan) and a biogas meter (MODEL 750 HP, LANDIS & GYR, Australia). The biogas composition was analyzed weekly with a multichannel portable gas analyzer (GFM406, Gas Data, Coventry, U.K.). Experimental results were statistically analyzed, and differences between groups were evaluated using either a 2-sample t-test or ANOVA, with a 95% confidence level applied for significance testing. Previous ML models will be used to compare against and provide insight into experimental results.

7.4 Results and Discussion

7.4.1 Pearson's Analysis for Linear Relationship Between Any Two Variables

Understanding the relationships between specific methane yield (SMY) and various input features is essential for optimizing biogas production from lignocellulosic biomass. This section provides a detailed exploration of how these variables influence SMY, offering valuable insights for optimizing anaerobic digestion processes. By understanding these correlations, we can identify strategies to enhance biogas production and improve the efficiency of lignocellulosic biomass conversion. In this section, we analyze these relationships using the Pearson correlation coefficient (PCC) matrix, as shown in Fig. 7.1. The PCC matrix and associated significance levels reveal strong linear correlations between SMY and several key variables: biomass size (Size), reactor volume (Vol), temperature (T), carbon content (C), nitrogen content (N), carbon-to-nitrogen ratio (CN), and lignin content (Lig) ($p < 0.01$). The p-values confirm the reliability of these correlations, providing a robust foundation for interpreting the results. The analysis identified both negative and positive correlations. Negative correlations were observed for Size, Vol, CN, and Lig, with coefficients of −0.21, −0.26, −0.27, and −0.25, respectively ($p < 0.01$). A negative correlation means that increasing these features reduces SMY. The opposite is true for positive correlations.

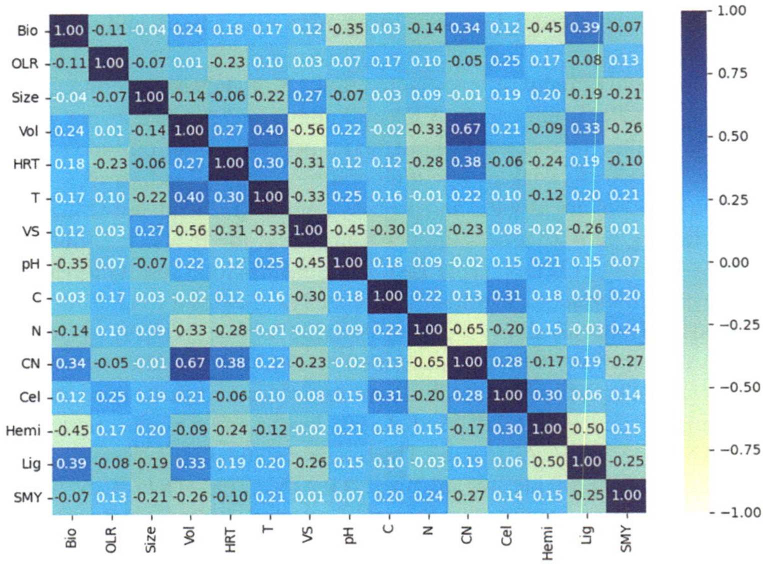

Fig. 7.1 Pearson correlation matrix between any two variables (*Source* with permission from Sonwai et al. 2023)

Negative Correlations:

1. Size: Smaller biomass particles result in increased SMY due to a greater surface area accessible to microbial degradation (Kaur 2022).
2. Vol: Larger reactors may reduce stirring efficiency or microorganism dispersion, lowering SMY.
3. CN: High C ratios indicate nitrogen deficiency, which inhibits microbial activity and reduces SMY (Neshat et al. 2017).
4. Lig: Lignin acts as a structural barrier, making decomposition difficult and decreasing SMY (Li et al. 2021).

Positive Correlations:

1. Temperature (T): A positive correlation of 0.21 indicates that higher temperatures enhance enzymatic activities and chemical reaction rates, boosting SMY (Neshat et al. 2017; Voelklein et al. 2016).
2. Carbon Content (C): A correlation of 0.20 suggests that carbon-rich feedstocks generally promote biogas production, although the nature of the carbon (e.g., recalcitrant lignin) must be considered (De Clercq et al. 2019; Djavan et al. 2020; Wang et al. 2021).
3. Nitrogen Content (N): The positive correlation (0.24) highlights the importance of nitrogen in maintaining microbial growth and achieving a balanced C ratio, leading to higher SMYs (Zhang et al. 2020).

Key Insights:

- The findings align with prior studies, showing that reduced biomass size, optimized C ratios, and controlled lignin content can significantly improve SMY.
- Carbon-rich biomass often suffers from nitrogen deficiencies, which can be mitigated by nitrogen supplementation for an optimal C ratio. This adjustment enhances biogas yield and stabilizes the system (Zhang et al. 2020).

7.4.2 Model Prediction Accuracy

In this section, Case 1 and Case 2 refer to two different datasets used to train and test machine learning (ML) models for predicting specific methane yield (SMY) from lignocellulosic biomass.

Case 1 includes all input features, providing a comprehensive dataset for the models to analyze. This case resulted in higher predictive accuracy, as evidenced by the higher R^2 values and lower root mean square error (RMSE) values. Case 2 excludes three critical biomass composition features: cellulose, hemicellulose, and lignin. This exclusion led to a noticeable drop in predictive accuracy, with the RF model's R^2 decreasing to 0.73.

The results from each machine learning (ML) algorithm are summarized in Fig. 7.2. When the models were tested for their predictive accuracy (measured by R^2), their results were higher for Case 1, which included all input features (Fig. 7.2a, c, and e), than for Case 2, which excluded cellulose, hemicellulose, and lignin (Fig. 7.2b, d and f). Among all the algorithms, RF emerged as the best for predicting specific methane yield (SMY), achieving the highest R^2 value of 0.85 in Case 1. However, predictive accuracy dropped in Case 2 ($R^2 = 0.73$), highlighting the importance of biomass composition—such as cellulose, hemicellulose, and lignin—in SMY predictions. This improvement with additional features was also evident in the RMSE values, which were lower for Case 1 than Case 2 (0.06 vs. 0.08).

RF has also been highlighted in previous studies as an effective algorithm for methane yield prediction. For instance, Cheon et al. (2022) and Park et al. (2021) found RF to be the best-performing model for predicting methane yields in industrial-scale anaerobic co-digestion, with R^2 values up to 0.88. Additionally, Long et al. (2021) reported an R^2 of 0.82 for methane yield predictions based on genomic data and operational parameters such as volatile fatty acids (VFAs), temperature, organic loading rates (OLRs), hydraulic retention time (HRT), and waste types. The RF accuracy in this chapter ($R^2 = 0.85$) aligns well with these findings.

However, more accurate models, such as those using recurrent neural networks (RNNs), have been reported. For example, RNNs achieved an impressive R^2 of 0.97, likely due to the use of a significantly larger dataset and input features derived from a single substrate (food waste). The comparison between Case 1 and Case 2 underscores the importance of biomass composition in modeling biogas production. Including features like cellulose, hemicellulose, and lignin improves the model's ability to capture the complex relationships influencing SMY. While RF performed

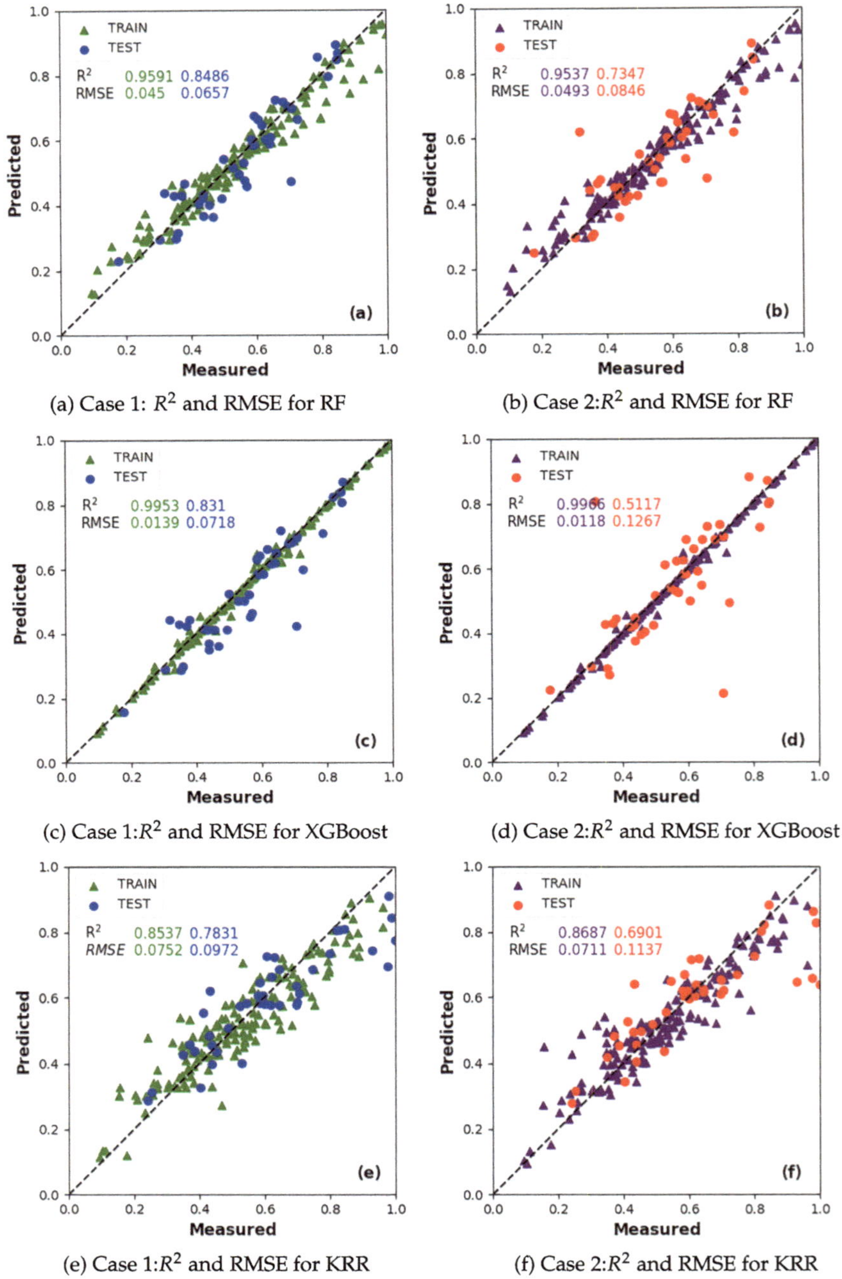

(a) Case 1: R^2 and RMSE for RF

(b) Case 2: R^2 and RMSE for RF

(c) Case 1: R^2 and RMSE for XGBoost

(d) Case 2: R^2 and RMSE for XGBoost

(e) Case 1: R^2 and RMSE for KRR

(f) Case 2: R^2 and RMSE for KRR

Fig. 7.2 Data distribution through each studied algorithm with R^2 and RMSE. Dotted lines show the trend of equality between predicted and the test values (*Source* with permission from Sonwai et al. 2023) well in both cases, its accuracy in Case 1 aligns with findings from previous studies, which have also identified RF as a reliable algorithm for methane yield prediction. However, more advanced models, such as recurrent neural networks (RNNs), have achieved even higher accuracy in some cases, particularly when using larger datasets and more specific input features.

7.4.3 Feature Evaluation and Interpretation

The factors that influence specific methane yield (SMY) were analyzed in this chapter using the plot importance method based on the Random Forest (RF) model, with corresponding Pearson correlation coefficients (PCCs) provided for comparison (Fig. 7.3). The analysis pinpointed cellulose and lignin as the most important components affecting how lignocellulosic biomass (LB) breaks down without oxygen, contributing 32% and 18% to the SMY predictions, respectively. Interestingly, cellulose had a greater impact on SMY than lignin, which goes against the common idea that lignin is the main factor because it can prevent microbes from breaking down biomass (Li et al. 2021; Neshat et al. 2017). Some past studies, like Liu et al. (2015) and Edwiges et al. (2018), have said that lignin is the most important thing for predicting SMY, while cellulose was often seen as less important or not important at all (Bayard et al. 2016; Nurk et al. 2016).

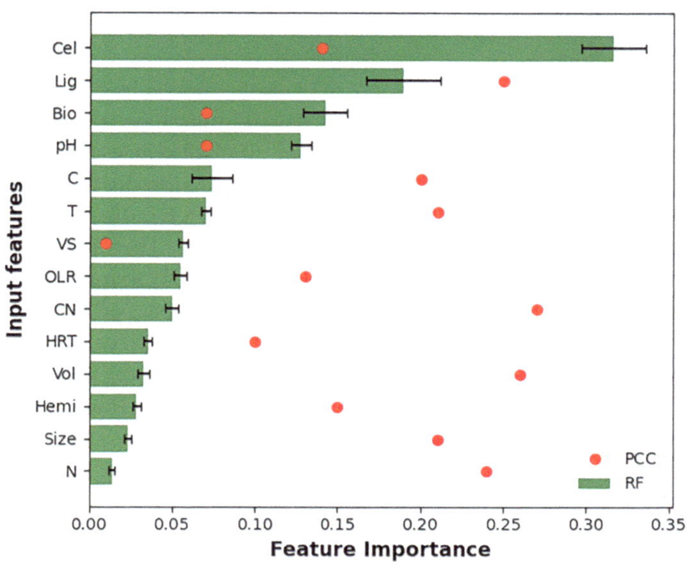

Fig. 7.3 Feature importance of SMY by RF model and the corresponding Pearson correlation coefficient (*Source* with permission from Sonwai et al. 2023)

Cellulose, which is known to break down easily, has had different results in previous research about its role in biogas production. Some studies have reported positive effects (Issah et al. 2020), while others have found no clear connection (Bayard et al. 2016; Nurk et al. 2016). These differences between this chapter and what others have found may come from different ways of analyzing data. Many earlier studies used statistical tools like principal component analysis (PCA), partial least squares (PLS), and multiple linear regression (MLR), but these aren't as good at handling the complex relationships that happen when breaking down materials without oxygen. The RF model, on the other hand, takes into account both simple and complex relationships, giving a more complete analysis (Zhu et al. 2019).

The PCCs calculated in this chapter also show how complex this is, with a lower connection for cellulose ($r = 0.14$, $p = 0.05$) compared to lignin ($r = -0.25$, $p < 0.01$) (Fig. 7.3). However, RF predictions showed that a high amount of cellulose can greatly increase SMY, as shown by similar yields from fresh *Pennisetum purpureum* × *Pennisetum typhoideum*, which has high cellulose, and other energy crops with less lignin (Sonwai et al. 2021). Also, results from pilot-scale reactors in Sect. 7.4.6 agreed with the RF model's predictions. The difference between how important something is and its PCC values shows that we need more advanced algorithms like RF to better understand the processes involved in breaking down LB without oxygen.

The amount of biomass was found to be another important thing that affects SMY. High SMYs were seen when biomass was broken down with other materials, especially when mixed with manures (Wall et al. 2014). Also, feedstocks with a lot of organic content were found to be important for getting the most biogas (De Clercq et al. 2019; Djavan et al. 2020; Wang et al. 2021). These findings suggest that the best SMYs can be reached with high LB amounts, as long as the biomass has low lignin and high cellulose.

When looking at how things are done, pH was the most important factor, having twice the impact as temperature. Keeping pH within the ideal range of 6.80–7.20 is important for keeping the process stable and efficient, since changes can cause volatile fatty acid (VFA) to build up and reduce biogas production (Neshat et al. 2017). While temperature also affects biogas production by affecting the microbes present, most studies work within ideal temperature ranges. Thermophilic conditions are usually better than mesophilic ones for better biogas conversion (Liu et al. 2022). Biomass particle size, however, didn't have much of an impact on SMY predictions, probably because most studies prepared biomass to the best sizes beforehand.

Overall, this analysis shows how important LB composition is—especially cellulose and lignin—and points out pH as the most important thing to consider for accurate SMY predictions.

7.4.4 Partial Dependence Plots (PDP)

In this chapter, we explore how six key factors—cellulose (Cel), lignin (Lig), biomass ratio (Bio), pH, carbon content (C), and temperature (T)—affect the specific methane yield (SMY) during anaerobic digestion. To visualize these effects, we used partial dependence plots (PDPs), which show how each factor influences SMY. The x-axis represents the distribution of data for each factor, while the y-axis shows its impact on SMY (Fig. 7.4).

Among these factors, cellulose had the most significant impact on SMY, with higher cellulose levels leading to increased methane production. This is because cellulose is a readily available source of energy for microbes during digestion. Conversely, lignin had a negative effect, as it can hinder microbial access to cellulose, slowing down the digestion process. This aligns with previous findings that lignin can limit hydrolysis, a critical step in anaerobic digestion (Kaur 2022).

The biomass ratio also played a crucial role in SMY predictions. The highest increase in SMY was observed when the biomass ratio was between 75% and 80% (Fig. 7.4c). However, co-digestion studies often suggest that a 1:1 biomass-to-manure ratio is optimal (Liu et al. 2015; Lianhua et al. 2020). This discrepancy is explored further in Sect. 7.4.5.

The remaining factors—pH, carbon content, and temperature—are essential for maintaining optimal conditions during anaerobic digestion. SMY increased with pH levels above 6.78, peaking at 7.60, which is within the recommended range for stability. For carbon content, SMY peaked at 42% before declining, possibly due to nitrogen deficiency at higher carbon levels. Temperature had the strongest influence between 35 °C and 55 °C, confirming the advantage of thermophilic conditions over mesophilic ones for biogas production.

To further understand how these factors interact, we conducted a two-way PDP analysis (Fig. 7.5). Notably, high SMYs were achieved when there was high cellulose ($\geq$42%) and low lignin ($\leq$27%) content. This suggests that sufficient cellulose can offset the negative effects of lignin. Another key interaction was observed between biomass ratio and organic loading rate (OLR), where high SMYs were predicted at moderate OLRs and high biomass ratios. Although higher OLRs typically lead to greater biogas production rates, excessive organic content can cause process inhibition and reduce gas yield (Guan et al. 2022).

Lastly, the interaction between pH and temperature was significant (Fig. 7.5c). High SMY was observed at temperatures between 35 °C and 55 °C combined with a pH range of 7.75–8.00, slightly above the conventional optimal range. This indicates that additional alkalinity may be needed to maintain optimal digester performance when using lignocellulosic biomass as feedstock. Both one-way and two-way PDP analyses provided valuable insights into how biomass characteristics and operating conditions affect anaerobic digestion performance.

7.4.5 Optimization of the Proportion of Co-digestion Substrates

In this chapter, we discuss how machine learning models, such as the Random Forest (RF) model, can help identify the optimal ratio of lignocellulosic biomass to manure

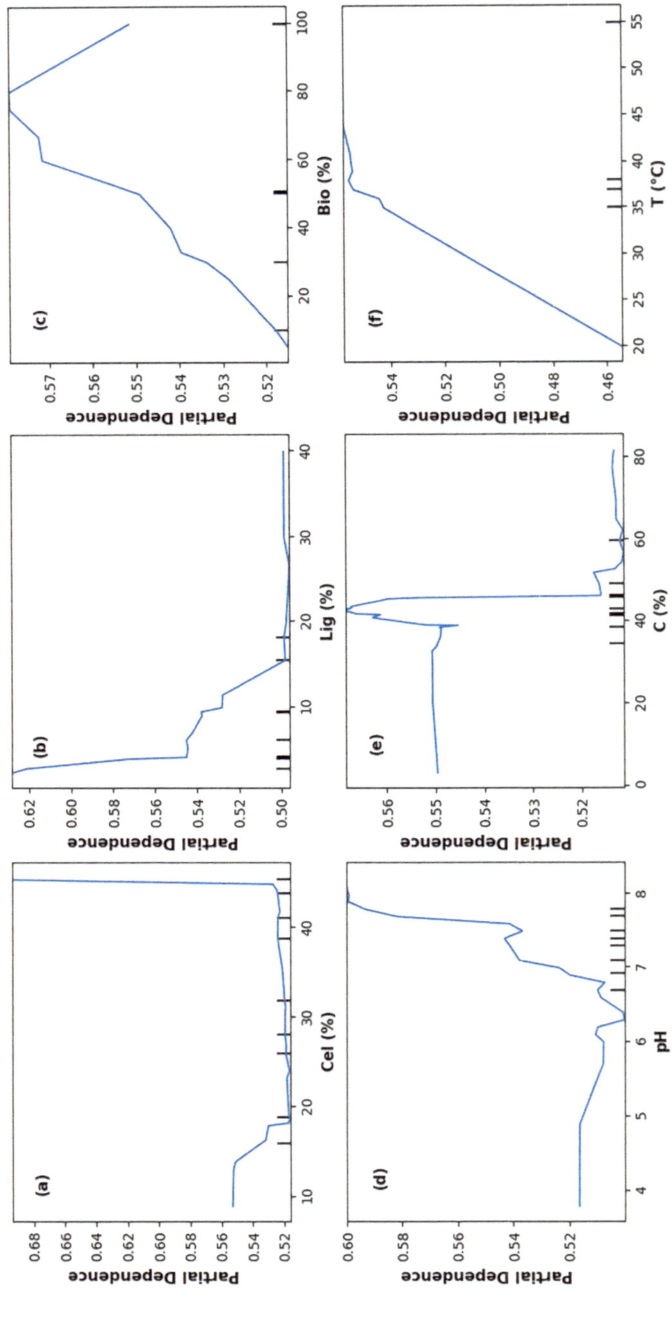

Fig. 7.4 One-way PDPs of the first six most important features by RF model (*Source* with permission from Sonwai et al. 2023)

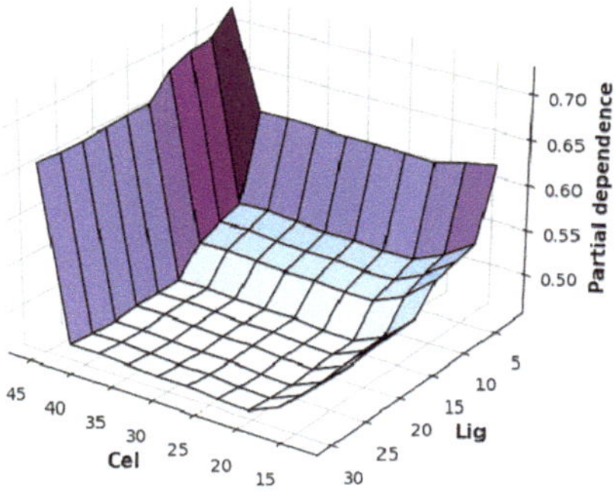

(a) Two way-PDP for cellulose and lignin

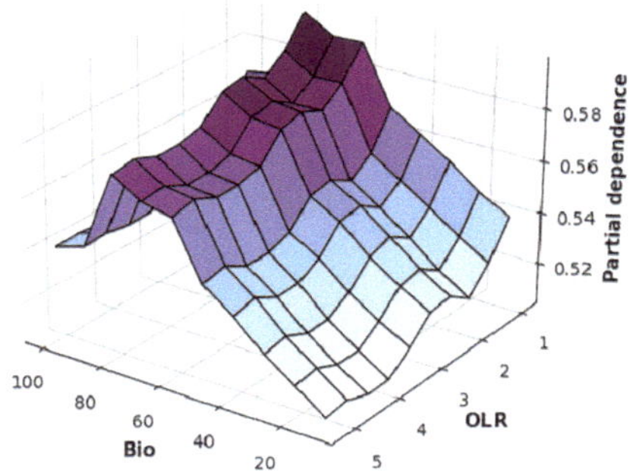

(b) Two way-PDP for biomass and OLR

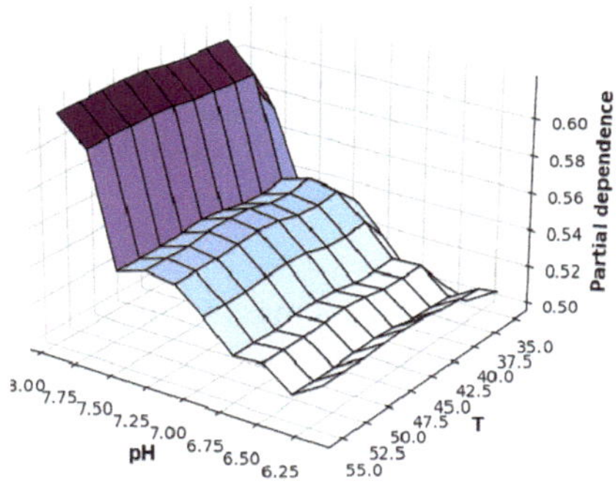

(c) Two way-PDP for pH and temperature

Fig. 7.5 Two-way PDPs of important features from RF predictions for SMY (*Source* with permission from Sonwai et al. 2023)

for maximizing specific methane yield (SMY) during anaerobic digestion. The RF model predicted the highest SMY of 178 mL CH_4 g VS^{-1} at a low organic loading rate (OLR) of 1.0 kg VS m^{-3} d^{-1} with an 80:20 biomass-to-manure ratio (Fig. 7.6). At lower OLRs, the digester can handle a higher biomass ratio because microorganisms require fewer nutrients to process smaller amounts of organic matter.

However, as OLRs increase to 2.0 and 2.5 kg VS m^{-3} d^{-1}, the optimal biomass ratio decreases to around 55%, resulting in slightly lower SMYs of 168 and 167 mL CH_4 g VS^{-1}, respectively. For even higher OLRs (3.0–4.0 kg VS m^{-3} d^{-1}), the optimal biomass ratio is further reduced to 50%, accompanied by a gradual decline in predicted SMYs. This reduction is due to the increased nutrient demand of microorganisms and the potential accumulation of volatile fatty acids (VFA) at higher OLRs, which can inhibit microbial activity and reduce biogas yield (Pan et al. 2022).

Previous studies have explored various optimal ratios for co-digesting lignocellulosic biomass and manure, ranging from 1:1 to 4:1 (Li et al. 2015; Lianhua et al.

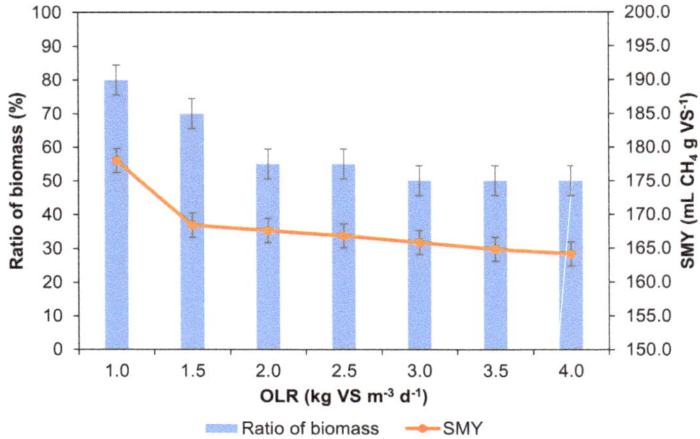

Fig. 7.6 Optimum biomass ratios at different OLRs with predicted SMY values. Ratio of biomass is presented as a percentage of biomass in the mixture of substrates, e.g., ratio of biomass = 80% means ratio of biomass to manure = 80:20, VS basis (*Source* with permission from Sonwai et al. 2023)

2020; Wall et al. 2014). The 1:1 biomass-to-manure ratio is often identified as the most effective, as seen in studies where rice straw was co-digested with cow or pig manure at a 50:50 ratio, yielding high methane production (Li et al. 2015). These findings emphasize the importance of balancing feedstock composition to optimize anaerobic digestion performance.

7.4.6 Comparison of Mono- and Co-digestion Experiments to Machine Learning predicted results

The study applied a Random Forest (RF) model to optimize the co-digestion of lignocellulosic biomass (LB) and cattle manure (CM) at various operating conditions, yielding critical insights into anaerobic digestion (AD) performance. Mono- and co-digestion trials were conducted under thermophilic conditions (55.0–55.6 °C) and near-neutral pH (7.24–7.46) across different organic loading rates (OLRs), with notable findings summarized below.

Mono-Digestion

1. Substrate Variability: The mono-digestion of giant juncao grass (GJG) outperformed that of rice straw (RS) and corn stover (CS) in terms of specific methane yield (SMY), particularly at lower OLRs (1.0–2.0 kg VS m^{-3} d^{-1}). GJG achieved the highest SMY (228 ± 14.5 mL CH_4 g VS^{-1}) due to its higher cellulose content, as supported by RF analysis. Conversely, RS and CS, with higher lignin and lower C:N ratios, exhibited inferior SMYs.

2. OLR Effects: At the highest OLR of 4.0 kg VS m^{-3} d^{-1}, GJG maintained superior performance with an SMY of 210 ± 10.7 mL CH_4 g VS^{-1}, which was both stable and significantly higher than those of RS and CS.

Co-Digestion

1. Improved Stability: Co-digestion of LB with CM (1:1 VS basis) enhanced process stability, as evidenced by lower volatile fatty acid (VFA) concentrations and VFA-to-alkalinity ratios at high OLRs (4.0 kg VS m^{-3} d^{-1}). This balance suggests better microbial activity and reduced risk of acidification.
2. Reduced SMYs: Co-digestion resulted in lower SMYs (130 ± 11.2, 121 ± 12.6 and 125 ± 14.6 mL CH_4 g VS^{-1} for GJG, RS, and CS, respectively) compared to mono-digestion. This aligns with previous reports indicating SMY reduction when manure is introduced as a co-substrate. However, similar SMYs across all substrates suggest that the co-digestion process diminishes the substrate-specific advantages seen in mono-digestion.
3. Model Validation: Experimental co-digestion SMYs were within 73.7–79.2% of the RF-predicted value (164 mL CH_4 g VS^{-1}), demonstrating the robustness of the model.

Key Insights and Implications

- Predictive Power of ML: The RF model effectively captured the complex interactions between substrate characteristics and operating conditions, highlighting the importance of cellulose over lignin content in achieving high SMYs.
- Optimal Conditions: The study validated the 1:1 LB-to-CM ratio (VS basis) as optimal under practical OLRs, emphasizing the balance of nutrients and microbial activity in co-digestion.
- Practical Considerations: While the developed ML model showed promise for optimizing lab-scale systems, further validation and potential modifications are required for full-scale applications. Additionally, exploring other manure types as co-substrates could expand the model's applicability (Table 7.4).

7.5 Conclusions

This chapter demonstrates the successful application of machine learning (ML) to predict and optimize specific methane yields (SMYs) from the anaerobic digestion (AD) of lignocellulosic biomass (LB) using continuous stirred-tank reactors (CSTRs). The study revealed several key insights that are valuable for improving biogas production processes.

- First, ML identified cellulose and lignin as the most influential features affecting SMY predictions. High cellulose content was found to significantly enhance SMY, while lignin had an inhibitory effect, making biomass with lower lignin content more favorable for biogas production. These findings were consistent with experimental data, confirming the reliability of the ML models.

Table 7.4 Performance parameters of each reactor at different studied conditions, from Sonwai et al. (2023) (with permission)

Condition	Reactors	OLR (kg VS m^{-3} d^{-1})	Temperature (°C)	pH	TS (mg L^{-1})	VS (mg L^{-1})	VFA (mg L^{-1})	Alk (mg L^{-1} as $CaCO_3$)	VFA/Alk	SMY (mLCH_4 g VS^{-1})
Mono-digestion	GR	1.0	55.0 ± 0.49^{a}	7.46 ± 0.08^{a}	40,607 ± 2,604^{b}	24,779 ± 1,923^{b}	528 ± 63.2^{b}	3,897 ± 276^{a}	0.14 ± 0.02^{c}	206 ± 10.1^{a}
	RR		55.2 ± 0.26^{a}	7.25 ± 0.13^{b}	51,579 ± 4,558^{a}	33,121 ± 4,152^{a}	645 ± 104^{b}	3,774 ± 357^{a}	0.17 ± 0.03^{b}	207 ± 17.0^{a}
	CR		55.1 ± 0.49^{a}	7.24 ± 0.11^{b}	54,393 ± 8,751^{a}	38,164 ± 7,238^{a}	832 ± 198^{a}	3,953 ± 671^{a}	0.21 ± 0.03^{a}	194 ± 9.21^{b}
	p-value		0.130	0.000	0.001	0.000	0.002	0.766	0.000	0.001
	GR	2.0	55.4 ± 0.35^{b}	7.38 ± 0.10^{a}	45,036 ± 4,048^{b}	28,329 ± 3,162^{b}	677 ± 75.0^{b}	4,445 ± 266^{a}	0.15 ± 0.01^{b}	228 ± 14.5^{a}
	RR		55.3 ± 0.28^{b}	7.26 ± 0.12^{b}	68,914 ± 7,325^{a}	46,079 ± 6,110^{a}	1,057 ± 151^{a}	4,174 ± 353^{a}	0.25 ± 0.03^{a}	207 ± 9.07^{b}
	CR		55.5 ± 0.34^{a}	7.33 ± 0.09^{a}	68,643 ± 3,685^{a}	51,164 ± 3,429^{a}	1,176 ± 144^{a}	4,250 ± 376^{a}	0.28 ± 0.03^{a}	181 ± 13.7^{c}
	p-value		0.005	0.000	0.000	0.000	0.000	0.320	0.000	0.000
	GR	3.0	55.5 ± 0.38^{ab}	7.28 ± 0.06^{b}	41,618 ± 6,260^{b}	25,045 ± 2,987^{b}	828 ± 104^{b}	4,909 ± 370^{a}	0.17 ± 0.03^{b}	214 ± 15.5^{a}
	RR		55.6 ± 0.34^{a}	7.29 ± 0.07^{b}	68,686 ± 11,351^{a}	45,600 ± 4,827^{a}	1,646 ± 145^{a}	5,006 ± 317^{a}	0.33 ± 0.02^{a}	172 ± 11.5^{b}
	CR		55.4 ± 0.38^{b}	7.33 ± 0.08^{a}	64,805 ± 8,921^{a}	50,614 ± 6,559^{a}	1,576 ± 136^{a}	4,798 ± 712^{a}	0.33 ± 0.04^{a}	166 ± 10.3^{b}
	p-value		0.038	0.000	0.000	0.000	0.000	0.623	0.000	0.000
	GR	4.0	55.3 ± 0.42^{a}	7.32 ± 0.03^{b}	53,600 ± 7,149^{b}	30,300 ± 5,342^{c}	1,463 ± 441^{b}	4,935 ± 243^{c}	0.29 ± 0.08^{b}	210 ± 10.7^{a}
	RR		55.4 ± 0.35^{a}	7.41 ± 0.03^{a}	74,469 ± 11,689^{a}	44,023 ± 6,297^{b}	2,135 ± 448^{a}	5,480 ± 202^{b}	0.39 ± 0.09^{a}	145 ± 8.82^{b}
	CR		55.4 ± 0.41^{a}	7.40 ± 0.02^{a}	72,746 ± 13,924^{a}	50,262 ± 6,598^{a}	2,211 ± 570^{a}	6,438 ± 734^{a}	0.34 ± 0.05^{ab}	140 ± 6.79^{b}
	p-value		0.419	0.000	0.000	0.000	0.001	0.000	0.009	0.000
Co-digestion	GR	4.0	55.5 ± 0.43^{a}	7.26 ± 0.03^{c}	70,118 ± 7,636^{b}	40,535 ± 4,715^{b}	1,257 ± 115^{b}	6,487 ± 524^{b}	0.20 ± 0.03^{c}	130 ± 11.2^{a}
	RR		55.3 ± 0.42^{b}	7.29 ± 0.05^{b}	108,424 ± 7,662^{a}	69,871 ± 5,924^{a}	2,359 ± 508^{a}	6,543 ± 441^{b}	0.36 ± 0.08^{a}	121 ± 12.6^{b}
	CR		55.5 ± 0.41^{a}	7.43 ± 0.03^{a}	107,812 ± 11,362^{a}	73,235 ± 8,067^{a}	2,479 ± 532^{a}	8,128 ± 551^{a}	0.30 ± 0.06^{b}	125 ± 14.6^{ab}
	p-value		0.008	0.000	0.000	0.000	0.000	0.000	0.000	0.042

Note Results are shown as Mean ± SD; Different superscripts in the same column indicate differences ($p < 0.05$)

- Second, the study determined the optimal biomass-to-manure ratio for co-digestion to be 1:1 (on a volatile solids basis) at organic loading rates (OLRs) of 3.0–4.0 kg VS m^{-3} d^{-1}. Experimental results showed that the achieved SMYs were approximately 79.2% of those predicted by the Random Forest (RF) model, highlighting the model's strong predictive accuracy.
- Beyond these specific findings, the study emphasized the broader potential of ML in advancing AD technology. By identifying critical factors and optimizing operating conditions, ML provides insights that are difficult to achieve through traditional experimental methods alone. This approach not only improves the efficiency of

biogas production but also reduces the time and resources required for process optimization.

While the study shows that high SMYs can be achieved with LB featuring high-cellulose and low-lignin content, further research is needed to adapt ML models for a wider range of feedstocks and industrial-scale applications. Expanding the scope of ML in this way will enhance its practical utility, making it a powerful tool for optimizing biogas production in real-world settings.

The practical implications of using Random Forest (RF) models in optimizing anaerobic digestion (AD) processes are significant, offering advancements in efficiency, cost-effectiveness, and scalability. These models address key challenges in AD, by leveraging data-driven insights to improve decision-making and operational outcomes.

Enhanced Prediction and Process Optimization RF models excel at predicting critical outputs like methane yield (SMY) and biogas composition by analyzing interactions between substrate properties (e.g., cellulose, lignin) and operational parameters (e.g., pH, temperature, and organic loading rate). For instance, RF's have the ability to identify optimal biomass-to-manure ratios, such as an 80:20 ratio at low organic loading rates (OLRs), which maximizes methane production while balancing nutrient availability. This predictive capability reduces reliance on trial-and-error approaches, enabling operators to fine-tune feedstock composition and loading rates for specific AD systems.

Real-Time Monitoring and Control RF models can act as "soft sensors" to monitor parameters like chemical oxygen demand (COD) in real time, facilitating automated process adjustments. For example, integrating RF with pH and temperature sensors allows systems to terminate digestion cycles once pollutant levels meet effluent standards, saving time and energy compared to fixed-time controls. This real-time adaptability enhances process stability and reduces the risk of acidification or microbial inhibition, particularly in co-digestion systems.

Cost and Energy Efficiency By optimizing feedstock selection and OLRs, RF models minimize resource waste and operational costs. They enable operators to prioritize high-cellulose substrates (e.g., guinea grass) over lignin-rich alternatives, improving methane yields by up to 228 mL CH_4 g VS^{-1} in mono-digestion. Additionally, RF-guided co-digestion balances nutrient availability, reducing volatile fatty acid accumulation and maintaining microbial activity even at high OLRs.

Scalability and Adaptability RF models demonstrate scalability from lab-scale experiments to full-scale industrial applications. This scalability is critical for diverse feedstocks and geographic locations, as RF accommodates site-specific variables such as digester volume and substrate composition.

Identification of Critical Variables Feature importance analysis in RF models highlights key factors influencing AD performance, such as cellulose content, pH, and temperature. For instance, cellulose's strong correlation with SMY underscores its role as a primary energy source for microbes, while lignin's inhibitory effects guide pretreatment strategies. This insight allows operators to prioritize adjustments to high-impact variables, streamlining process optimization.

Future Applications While RF models have proven effective in lab settings, further validation is needed for full-scale systems. Expanding these models to include diverse manure types and integrating them with advanced optimization algorithms (e.g., meta-heuristic methods) could enhance their predictive accuracy and applicability. Such advancements would reinforce AD's role in sustainable waste management and renewable energy production, aligning with global carbon neutrality goals.

In summary, RF models transform AD optimization by providing data-driven solutions for feedstock selection, process control, and resource efficiency, making them indispensable tools for advancing the circular bioeconomy initiatives.

References

Almomani F (2020) Prediction of biogas production from chemically treated co-digested agricultural waste using artificial neural network. Fuel 280:118573. ISSN 0016-2361. https://doi.org/10.1016/j.fuel.2020.118573. https://www.sciencedirect.com/science/article/pii/S0016236120315696

Andrade Cruz I, Chuenchart W, Long F, Surendra K, Renata Santos Andrade L, Bilal M, Liu H, Tavares Figueiredo R, Khanal SK, Fernando Romanholo Ferreira L (2022) Application of machine learning in anaerobic digestion: perspectives and challenges. Bioresour Technol 345:126433. ISSN 0960-8524. https://doi.org/10.1016/j.biortech.2021.126433. https://www.sciencedirect.com/science/article/pii/S0960852421017752

Bayard R, Liu X, Benbelkacem H, Buffiere P, Gourdon R (2016) Can biomethane potential (bmp) be predicted from other variables such as biochemical composition in lignocellulosic biomass and related organic residues? BioEnergy Res 9(2):610–623. ISSN 1939-1242. https://doi.org/10.1007/s12155-015-9701-3

Cheon A, Sung J, Jun H, Jang H, Kim M, Park J (2022) Application of various machine learning models for process stability of bio-electrochemical anaerobic digestion. Processes 10(1). ISSN 2227-9717. https://doi.org/10.3390/pr10010158. https://www.mdpi.com/2227-9717/10/1/158

De Clercq D, Jalota D, Shang R, Ni K, Zhang Z, Khan A, Wen Z, Caicedo L, Yuan K (2019) Machine learning powered software for accurate prediction of biogas production: a case study on industrial-scale chinese production data. J Cleaner Prod 218:390–399. ISSN 0959-6526. https://doi.org/10.1016/j.jclepro.2019.01.031. https://www.sciencedirect.com/science/article/pii/S095965261930037X

DiLallo R, Albertson OE (1961) Volatile acids by direct titration. J (Water Poll Control Feder) 356–365

Djavan D, Zongguo W, Fan F, Luis C, Kai Y, Ruoxi S (2020) Interpretable machine learning for predicting biomethane production in industrial-scale anaerobic co-digestion. Sci Total Envir 712. https://doi.org/10.1016/j.scitotenv.2019.134574

Edwiges T, Frare L, Mayer B, Lins L, Mi Triolo J, Flotats X, de Mendonca Costa MSS (2018) Influence of chemical composition on biochemical methane potential of fruit and vegetable waste. Waste Manag 71:618–625. ISSN 0956-053X. https://doi.org/10.1016/j.wasman.2017.05.030. https://www.sciencedirect.com/science/article/pii/S0956053X17303574

FitzGerald JA, Wall DM, Jackson SA, Murphy JD, Dobson AD (2019) Trace element supplementation is associated with increases in fermenting bacteria in biogas mono-digestion of grass silage. Renew Energy 138:980–986. ISSN 0960-1481. https://doi.org/10.1016/j.renene.2019.02.051. https://www.sciencedirect.com/science/article/pii/S0960148119302058

Guan R, Yuan H, Yuan S, Yan B, Zuo X, Chen X, Li X (2022) Current development and perspectives of anaerobic bioconversion of crop stalks to biogas: a review. Bioresour Technol 349:126615. ISSN 0960-8524. https://doi.org/10.1016/j.biortech.2021.126615. https://www.sciencedirect.com/science/article/pii/S096085242101957X

Gupta R, Zhang L, Hou J, Zhang Z, Liu H, You S, Sik Ok Y, Li W (2023) Review of explainable machine learning for anaerobic digestion. Bioresour Technol 369:128468. ISSN 0960-8524. https://doi.org/10.1016/j.biortech.2022.128468. https://www.sciencedirect.com/science/article/pii/S0960852422018016

In-Kwon Y, A JR (2000) A new family of power transformations to improve normality or symmetry. Biometrika 87(4):954–959. ISSN 0006-3444. https://doi.org/10.1093/biomet/87.4.954

Issah A-A, Kabera T, Kemausuor F (2020) Biogas optimisation processes and effluent quality: a review. Biomass Bioenergy 133:105449. ISSN 0961-9534. https://doi.org/10.1016/j.biombioe.2019.105449. https://www.sciencedirect.com/science/article/pii/S0961953419303988

Karki R, Chuenchart W, Surendra K, Shrestha S, Raskin L, Sung S, Hashimoto A, Kumar Khanal S (2021) Anaerobic co-digestion: Current status and perspectives. Bioresour Technol 330:125001. ISSN 0960-8524. https://doi.org/10.1016/j.biortech.2021.125001. https://www.sciencedirect.com/science/article/pii/S0960852421003400

Kasinath A, Fudala-Ksiazek S, Szopinska M, Bylinski H, Artichowicz W, Remiszewska-Skwarek A, Luczkiewicz A (2021) Biomass in biogas production: pretreatment and codigestion. Renew Sustain Energy Rev 150:111509. ISSN 1364-0321. https://doi.org/10.1016/j.rser.2021.111509. https://www.sciencedirect.com/science/article/pii/S1364032121007887

Kaur M (2022) Effect of particle size on enhancement of biogas production from crop residue. Mater Today: Proc 57:1950–1954. ISSN 2214-7853. https://doi.org/10.1016/j.matpr.2022.03.292. https://www.sciencedirect.com/science/article/pii/S2214785322016303. International Chemical Engineering Conference 2021 (100 Glorious Years of Chemical Engineering I& Technology)

Khan M, Chuenchart W, Surendra K, Kumar Khanal S (2023) Applications of artificial intelligence in anaerobic co-digestion: recent advances and prospects. Bioresour Technol 370:128501. ISSN 0960-8524. https://doi.org/10.1016/j.biortech.2022.128501. https://www.sciencedirect.com/science/article/pii/S096085242201834X

Li D, Liu S, Mi L, Li Z, Yuan Y, Yan Z, Liu X (2015) Effects of feedstock ratio and organic loading rate on the anaerobic mesophilic co-digestion of rice straw and pig manure. Bioresour Technol 187:120–127. ISSN 0960-8524. https://doi.org/10.1016/j.biortech.2015.03.040. https://www.sciencedirect.com/science/article/pii/S0960852415003703

Li H, Ke L, Chen Z, Feng G, Xia D, Wang Y, Zheng Y, Li Q (2016) Estimating the fates of c and n in various anaerobic codigestions of manure and lignocellulosic biomass based on artificial neural networks. Energy Fuels 30(11):9490–9501. ISSN 0887-0624. https://doi.org/10.1021/acs.energyfuels.6b01883

Li P, Liu D, Pei Z, Zhao L, Shi F, Yao Z, Li W, Sun Y, Wang S, Yu Q, Chen L, Liu J (2021) Evaluation of lignin inhibition in anaerobic digestion from the perspective of reducing the hydrolysis rate of holocellulose. Bioresour Technol 333:125204. ISSN 0960-8524. https://doi.org/10.1016/j.biortech.2021.125204. https://www.sciencedirect.com/science/article/pii/S0960852421005435

Li W, Khalid H, Zhu Z, Zhang R, Liu G, Chen C, Thorin E (2018) Methane production through anaerobic digestion: participation and digestion characteristics of cellulose, hemicellulose and lignin. Appl Energy 226:1219–1228. ISSN 0306-2619. https://doi.org/10.1016/j.apenergy.2018.05.055. https://www.sciencedirect.com/science/article/pii/S0306261918307657

Li Y, Zhang R, He Y, Zhang C, Liu X, Chen C, Liu G (2014) Anaerobic co-digestion of chicken manure and corn stover in batch and continuously stirred tank reactor (cstr). Bioresour Technol 156:342–347. ISSN 0960-8524. https://doi.org/10.1016/j.biortech.2014.01.054. https://www.sciencedirect.com/science/article/pii/S0960852414000790

Lianhua L, Shuibin H, Yongming S, Xihui K, Junfeng J, Zhenhong Y, Dingfa L (2020) Anaerobic co-digestion of pennisetum hybrid and pig manure: a comparative study of performance and microbial community at different mixture ratio and organic loading rate. Chemosphere

247:125871. ISSN 0045-6535. https://doi.org/10.1016/j.chemosphere.2020.125871. https://www.sciencedirect.com/science/article/pii/S004565352030062X

Liu X, Bayard R, Benbelkacem H, Buffiere P, Gourdon R (2015) Evaluation of the correlations between biodegradability of lignocellulosic feedstocks in anaerobic digestion process and their biochemical characteristics. Biomass Bioenergy 81:534–543. ISSN 0961-9534. https://doi.org/10.1016/j.biombioe.2015.06.021. https://www.sciencedirect.com/science/article/pii/S0961953415300349

Liu Y, Wang T, Xing Z, Ma Y, Nan F, Pan L, Chen J (2022) Anaerobic co-digestion of chinese cabbage waste and cow manure at mesophilic and thermophilic temperatures: digestion performance, microbial community, and biogas slurry fertility. Bioresour Technol 363:127976. ISSN 0960-8524. https://doi.org/10.1016/j.biortech.2022.127976. https://www.sciencedirect.com/science/article/pii/S0960852422013098

Long F, Wang L, Cai W, Lesnik K, Liu H (2021) Predicting the performance of anaerobic digestion using machine learning algorithms and genomic data. Water Res 199:117182. ISSN 0043-1354. https://doi.org/10.1016/j.watres.2021.117182. https://www.sciencedirect.com/science/article/pii/S0043135421003808

Meena M, Shubham S, Paritosh K, Pareek N, Vivekanand V (2021) Production of biofuels from biomass: predicting the energy employing artificial intelligence modelling. Bioresour Technol 340:125642. ISSN 0960-8524. https://doi.org/10.1016/j.biortech.2021.125642. https://www.sciencedirect.com/science/article/pii/S0960852421009834

Neshat SA, Mohammadi M, Najafpour GD, Lahijani P (2017) Anaerobic co-digestion of animal manures and lignocellulosic residues as a potent approach for sustainable biogas production. Renew Sustain Energy Rev 79:308–322. ISSN 1364-0321. https://doi.org/10.1016/j.rser.2017.05.137. https://www.sciencedirect.com/science/article/pii/S1364032117307694

Nurk L, Buhle L, Wachendorf M (2016) Degradation of fibre and non-fibre fractions during anaerobic digestion in silages of maize, sunflower and sorghum-sudangrass of different maturities. BioEnergy Res 9(3):720–730. ISSN 1939-1242. https://doi.org/10.1007/s12155-016-9717-3

Pan S, Zabed HM, Li Z, Qi X, Wei Y (2022) Enrichment and balancing of nutrients for improved methane production using three compositionally different agro-livestock wastes: process performance and microbial community analysis. Bioresour Technol 357:127360. ISSN 0960-8524. https://doi.org/10.1016/j.biortech.2022.127360. https://www.sciencedirect.com/science/article/pii/S0960852422006897

Park J-G, Jun H-B, Heo T-Y (2021) Retraining prior state performances of anaerobic digestion improves prediction accuracy of methane yield in various machine learning models. Appl Energy 298:117250. ISSN 0306-2619. https://doi.org/10.1016/j.apenergy.2021.117250. https://www.sciencedirect.com/science/article/pii/S030626192100670X

Pedregosa F, Varoquaux G, Gramfort A, Michel V, Thirion B, Grisel O, Blondel M, Prettenhofer P, Weiss R, Dubourg V et al (2011) Scikit-learn: Machine learning in python. J Mach Learn Res 12:2825–2830

Phromphithak S, Onsree T, Tippayawong N (2021) Machine learning prediction of cellulose-rich materials from biomass pretreatment with ionic liquid solvents. Bioresour Technol 323:124642. ISSN 0960-8524. https://doi.org/10.1016/j.biortech.2020.124642. https://www.sciencedirect.com/science/article/pii/S0960852420319167

Potter C, Klooster S, Genovese V, (2012) Net primary production of terrestrial ecosystems from 2000 to 2009. Clim Chan 115(2), 365–378. ISSN 1573-1480. https://doi.org/10.1007/s10584-012-0460-2

Ram A, Saurabh T, Kumar AA, Pankaj K (2019) Population-urbanization-energy nexus: a review. Resources 8(3). ISSN 2079-9276. https://doi.org/10.3390/resources8030136. https://www.mdpi.com/2079-9276/8/3/136

Sonwai A, Pholchan P, Pholchan MK, Pardang P, Nuntaphan A, Juangjandee P, Totarat N, Tippayawong N (2021) Biogas production from high solids digestion of pennisetum purpureum x pennisetum typhoideum: suitable conditions and microbial communities. J Environ Manag

299:113570. ISSN 0301-4797. https://doi.org/10.1016/j.jenvman.2021.113570. https://www.sciencedirect.com/science/article/pii/S0301479721016327

Sonwai A, Pholchan P, Tippayawong N (2023) Machine learning approach for determining and optimizing influential factors of biogas production from lignocellulosic biomass. Bioresour Technol 383:129235. ISSN 0960-8524. https://doi.org/10.1016/j.biortech.2023.129235. https://www.sciencedirect.com/science/article/pii/S0960852423006612

Thamsiriroj T, Nizami A, Murphy J (2012) Why does mono-digestion of grass silage fail in long term operation? Appl Energy 95:64–76. ISSN 0306-2619

Voelklein M, Rusmanis D, Murphy J (2016) Increased loading rates and specific methane yields facilitated by digesting grass silage at thermophilic rather than mesophilic temperatures. Bioresour Technol 216:486–493. ISSN 0960-8524. https://doi.org/10.1016/j.biortech.2016.05.109. https://www.sciencedirect.com/science/article/pii/S0960852416307684

Wall DM, Allen E, Straccialini B, Kiely PO, Murphy JD (2014) Optimisation of digester performance with increasing organic loading rate for mono- and co-digestion of grass silage and dairy slurry. Bioresour Technol 173:422–428. ISSN 0960-8524. https://doi.org/10.1016/j.biortech.2014.09.126. https://www.sciencedirect.com/science/article/pii/S0960852414013856

Wang L, Long F, Liao W, Liu H (2020) Prediction of anaerobic digestion performance and identification of critical operational parameters using machine learning algorithms. Bioresour Technol 298:122495. ISSN 0960-8524. https://doi.org/10.1016/j.biortech.2019.122495. https://www.sciencedirect.com/science/article/pii/S0960852419317250

Wang Y, Huntington T, Scown CD (2021) Tree-based automated machine learning to predict biogas production for anaerobic co-digestion of organic waste. ACS Sustainable Chem Eng 9(38):12990–13000. https://doi.org/10.1021/acssuschemeng.1c04612

Wang Z, Peng X, Xia A, Shah AA, Yan H, Huang Y, Zhu X, Zhu X, Liao Q (2023) Comparison of machine learning methods for predicting the methane production from anaerobic digestion of lignocellulosic biomass. Energy 263:125883. ISSN 0360-5442. https://doi.org/10.1016/j.energy.2022.125883. https://www.sciencedirect.com/science/article/pii/S0360544222027694

Zhang M, Wang Z, Zhang X, Qian X, Shen G (2020) Biogas and quality fertilizer production from dry anaerobic digestion of rice straw with nitrogen addition. Bioresour Technol Rep 11:100509. ISSN 2589-014X. https://doi.org/10.1016/j.biteb.2020.100509. https://www.sciencedirect.com/science/article/pii/S2589014X20301304

Zhu X, Li Y, Wang X (2019) Machine learning prediction of biochar yield and carbon contents in biochar based on biomass characteristics and pyrolysis conditions. Bioresour Technol 288:121527. ISSN 0960-8524. https://doi.org/10.1016/j.biortech.2019.121527. https://www.sciencedirect.com/science/article/pii/S0960852419307576

Chapter 8
Hydrothermal Carbonization

Abstract Hydrothermal carbonization is a process that converts biomass into a form of charcoal, offering an environmentally friendly way to reduce our reliance on natural resources and lower CO_2 emissions that contribute to global warming. However, understanding this process can be difficult due to its complexity and the many variables involved. To simplify this, this chapter uses a powerful machine learning tool called the extreme gradient boosting to predict important characteristics of the resulting hydrochar. These characteristics include how much hydrochar is produced, its energy content, ash content, and its composition of carbon, hydrogen, and oxygen. By analyzing a large dataset with over 1,000 data points and 16 different input factors, the predictions were very accurate, with strong correlations between the predicted and actual values. Furthermore, this chapter examines how different input factors affect these characteristics using advanced methods like Shapley values and SHAP dependence plots. This analysis provides valuable insights into the process and showcases an innovative approach, contributing significantly to the field.

Keywords Biomass conversion · Clean energy · Hydrochar · Machine learning

8.1 Introduction

Over the years, Earth's natural energy resources have been declining due to human consumption patterns, exacerbating global warming and disrupting the planet's natural balance (Katongtung et al. 2022b, 2024a). One promising option is energy derived from biomass or biofuels, which can be produced through thermal transformation processes (Katongtung et al. 2023).

One notable product of these processes is hydrochar, a carbon-rich material created through hydrothermal carbonization (HTC). This process involves heating biomass with minimal or no oxygen present. Hydrochar is used primarily to store carbon in soil and improve soil quality. It can capture up to 50% of the biomass's carbon—significantly more than the 3% achieved through biomass burning or the 20% (max) from natural decomposition over a 5–10 year period. The yield varies based on the type of biomass used (Reza et al. 2014; Al Afif et al. 2024).

N. Tippayawong et al., *Machine Learning Applications for Bioenergy Conversion*, Green Energy and Technology, https://doi.org/10.1007/978-981-96-8086-3_8

HTC relies on heat and water as catalysts to convert biomass at high temperatures (180–300 °C) and pressures (1.5–2.5 MPa) over 1–7 h. The process produces hydrochar along with a liquid byproduct (Rasaq et al. 2024). After filtration and drying, the hydrochar can be used as biofuel or as a soil amendment, potentially reducing air and water pollution (Wilk et al. 2024; Liu et al. 2024a).

Understanding the chemical reactions and factors influencing HTC can be complex. To address this, machine learning (ML) techniques are increasingly used to analyze and predict outcomes (Zhu et al. 2023; Mahmoud et al. 2024). ML has been effectively applied to related processes such as torrefaction, pyrolysis, and hydrothermal liquefaction (Katongtung et al. 2024b, 2022a). This chapter focuses on the use of Extreme Gradient Boosting (XGB), a highly accurate ML algorithm (Sukpancharoen et al. 2023; Janpoom et al. 2023).

There have been previous studies applying ML to HTC processes. For example, Liu et al. (2024b) used ML to optimize HTC parameters and experimentally validated their results using 333 data points across four ML models. Their study predicted outputs such as yield, ash content, carbon content, higher heating value (HHV), and energy recovery (ER), achieving a maximum R^2 value of 0.927.

Similarly, Zhang et al. (2023) provided an overview of ML modeling in HTC, summarizing results from various studies. Li et al. (2020) applied ML to predict hydrochar properties from wet waste using 248 data points and the SVR model, achieving an R^2 value of 0.92. Li et al. (2018) studied how feedstock properties and process conditions affected hydrochar yield, carbon content, and energy content, using 622 data points and achieving an R^2 value of 0.95 through RF modeling. Other notable studies include those by Castro Garcia et al. (2022), Mu et al. (2022), Djandja et al. (2021, 2022), each of which applied ML to predict key characteristics of hydrochar, with R^2 values ranging from 0.94 to 0.99.

While these studies have demonstrated the potential of ML in HTC, many were limited by small datasets, narrow predictive outputs, or outdated algorithms. In this chapter, larger datasets, broader output predictions, and modern ML algorithms are employed. The aim is to predict various hydrochar properties, including yield, calorific value, ash content, and elemental composition (carbon, hydrogen, and oxygen). To provide deeper insights, tools such as Shapley values and SHAP dependence plots are used to explore the relationships between critical variables influencing HTC reactions.

8.2 Materials and Methods

8.2.1 Hydrothermal Carbonization

Hydrothermal carbonization (HTC) is a process that uses water as a medium to break down biomass under high pressure and moderate temperatures. During HTC, water becomes subcritical, meaning its properties change significantly, allowing it to facilitate chemical reactions more effectively. This occurs at pressures typically between 1.5 and 2.5 MPa and temperatures ranging from 180 to 300 °C. In these conditions, water dissociates into hydronium ions (H_3O_+) and hydroxide ions (OH_-), which are crucial for the breakdown of biomass (Taglieri et al. 2024).

One of the key advantages of HTC is that it does not require biomass to be pre-dried, making it ideal for materials with high-moisture content, such as aquatic plants and organic waste. The process can accommodate a range of temperatures and pressures, leading to different reactions. For example, hydrolysis occurs at temperatures above 150 °C, breaking down complex molecules like cellulose, hemicellulose, and proteins into simpler ones. At higher temperatures and pressures, biomass can undergo liquefaction or form oil-like substances (Hansted et al. 2024).

The HTC process involves several key chemical reactions: hydrolysis, dehydration, decarboxylation, aromatization, and recondensation. These reactions transform biomass into hydrochar, which generally has lower volatile matter, higher fixed carbon content, reduced ash and moisture, and enhanced energy density and porosity (Chen et al. 2022). The resulting hydrochar can be used as a biofuel or as a soil amendment, offering potential benefits for reducing air and water pollution.

- Hydrolysis breaks down the chemical structure of biomass by cleaving ester and ether bonds with water molecules.
- Dehydration removes water from the biomass, initiating decomposition.
- Decarboxylation eliminates CO_2 by decomposing carboxyl groups in the biomass.
- Aromatization, which occurs alongside dehydration and decarboxylation, involves the formation of double bonds between carbon atoms or between carbon and oxygen. This process replaces hydroxyl and carboxyl groups, producing acids, aldehydes, and phenols.
- Recondensation follows the decomposition of biomass, leading to the formation of hydrochar. This step proceeds gradually, giving hydrochar its stable structure and properties (de Mora et al. 2024).

These reactions collectively transform high-moisture biomass into a solid product that is denser in energy and better suited for use as fuel or soil amendment.

8.2.2 Data Collection and ML Model

This chapter analyzes data from 1,200 studies, with 133 contributing directly to the analysis. The selected dataset (Dataset 1) includes 1,070 entries and focuses on 16 input variables—such as elemental properties and processing conditions—and six output variables. These variables are visualized in Figs. 8.1 and 8.2, with full details provided in Table 8.1.

To understand patterns in the data, box plots were used. These plots summarize data distributions using five key values: the smallest observation, the 25th percentile, the median, the 75th percentile, and the largest observation. They help identify central trends, variability, and outliers, making them particularly useful for comparing groups.

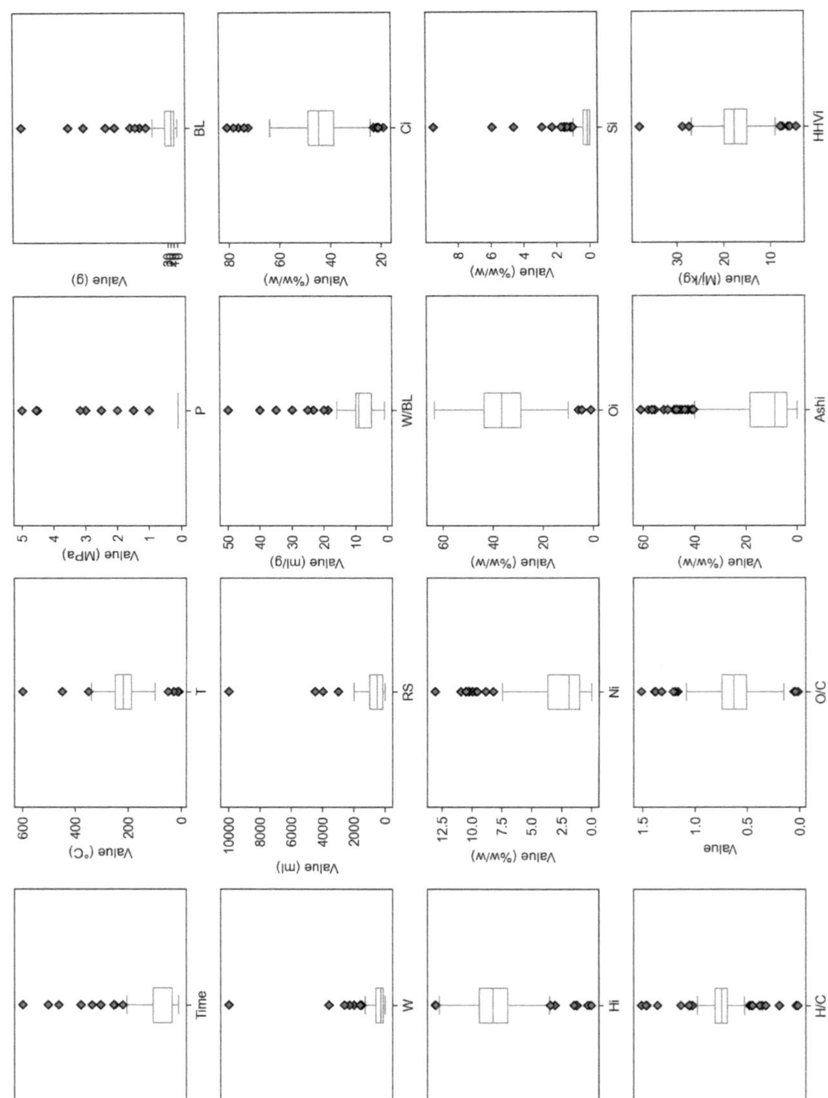

Fig. 8.1 Box plot of the distribution of the 16 input features, from Katongtung et al. (2024c) (with permission)

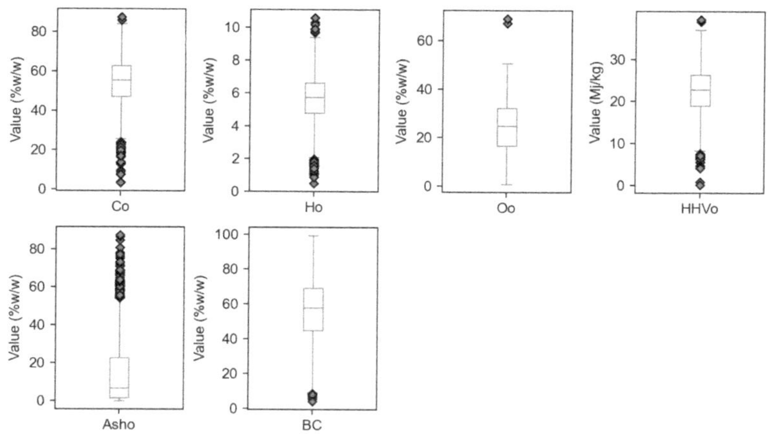

Fig. 8.2 Box plot of the distribution of the 6 target output features, from Katongtung et al. (2024c) (with permission)

The analysis relies on Extreme Gradient Boosting (XGB), a machine learning method known for its accuracy and efficiency. XGB excels at handling structured data and offers several advantages:

1. Speed: Its optimized design allows quick processing even with large datasets.
2. Overfitting prevention: Built-in techniques refine the model to avoid over-reliance on noisy data.
3. Interpretability: The algorithm highlights which factors most strongly influence predictions.
4. Flexibility: It works seamlessly with missing or incomplete data.

To ensure reliable results, the dataset was evaluated using tenfold cross-validation. This technique divides the data into 10 subsets, repeatedly training the model on nine subsets and testing it on the 10th. This approach minimizes bias and provides a realistic assessment of the model's performance (Peco Chacon et al. 2023). By combining thorough data preparation, advanced machine learning, and careful validation, this chapter demonstrates how modern computational tools can deepen our understanding of hydrothermal carbonization processes.

To evaluate the model's accuracy, two key metrics were used: the coefficient of determination (R^2) and the root mean square error (RMSE). The R^2 value measures how well the input features explain variations in the output features. For example, an R^2 of 0.90 means 90% of the changes in the output can be linked to the input variables. The RMSE, on the other hand, calculates the average difference between predicted and actual values, acting like a "prediction error meter." A smaller RMSE indicates better accuracy. For easier comparison across datasets, the normalized root mean square error (NRMSE) was also calculated, adjusting the RMSE to a standardized scale.

Higher R^2 values and lower RMSE values signal stronger predictive performance. To ensure the model was both accurate and reliable, its settings (called hyperparame-

Table 8.1 Set of input features/variables, from Katongtung et al. (2024c) (with permission)

Variable	Abbreviation	Value	Mean	SD
Input feature				
Feedstock characteristic (on dry basis)				
Elemental property				
Carbon content (% w/w)	C	19–80.7	44.5	8
Hydrogen content (% w/w)	Hi	0.9–8.9	5.9	1.3
Nitrogen content (% w/w)	Ni	0–13.1	1.9	2.7
Oxygen content (% w/w)	Oi	1.1–63.3	36.6	10.7
Sulfur content (% w/w)	Si	0–9.5	0.1	0.9
H-to-C atomic ratio (-)	H/C	0.1–3.2	1.6	0.3
O-to-C atomic ratio (-)	O/C	0–1.5	0.6	0.3
Ash (% w/w)	Ai	0–60.9	8.65	14.5
Higher heating value (MJ/kg)	HHVi	4.5–37.8	17.7	3.7
Operating condition				
Residence time (min)	Time	0–1440	60	189.7
Temperature (°C)	T	50–600	220	40.9
(Initial) pressure (MPa)	P	0.1–5	0.1	0.6
Reactor size (mL)	RS	10–10000	500	1224.9
Biomass loading (g)	BL	0.3–500	20	67.6
Water (mL)	W	3.5–5000	150	519.5
Water-to-BL ratio (mL/g)	W/BL	1–50	9	6.3
Output target				
Hydrochar yield (% w/w)	BC	4.1–99.2	57.9	18.2
Higher heating value (MJ/kg)	HHVo	0.1–39.4	22.8	5.7
Carbon content (% w/w)	Co	3.4–87.4	55.5	13.3
Hydrogen content (% w/w)	Ho	0.5–10.5	5.7	1.5
Oxygen content (% w/w)	Oo	0.8–68.9	24.8	10.6
Ash (% w/w)	Ao	0–87.5	6.8	18.6

ters) were carefully adjusted using a systematic search method paired with a tenfold cross-validation approach. This process helps avoid overfitting and ensures the model works well with new, unseen data. The finalized settings for the XGB algorithm are listed in Table 8.2.

8.2.3 *Shapley Values and SHAP Dependence Plot*

The Shapley value offers a way to understand how individual features influence a model's prediction for a specific data point. By adding up the Shapley values for all features, researchers can determine how much a prediction deviates from the dataset's

Table 8.2 The hyperparameters for 6 target outputs for XGB algorithm from Katongtung et al. (2024c) (with permission)

ML Algorithm	Hyperparameters	Optimized Value					
		BC	HHV	A	O	H	C
XGB	"subsample"	0.8	0.8	0.8	0.8	0.8	0.8
	"objective"	reg	reg	reg	reg	reg	reg
	"n_estimators"	850	850	850	850	850	850
	"max_depth"	90	90	90	90	90	90
	"learning_rate"	0.08	0.08	0.08	0.08	0.08	0.08
	"colsample_bytree"	0.9	0.9	0.9	0.9	0.9	0.9
	"min_child_weight"	7	7	7	7	7	7
	"colsample_bylevel"	0.95	0.95	0.95	0.95	0.95	0.95
	"colsample_bynode"	0.4	0.4	0.4	0.4	0.4	0.4
	"gamma"	2×10^{-5}	2×10^{-5}	2×10^{-5}	2×10^{-5}	2×10^{-5}	2×10^{-5}

Note Other unlisted hyperparameters were kept at default values. [a] The time (in ms) is spent in each algorithm to process both training and testing tasks in tenfold cross-validation

average prediction. This method provides a detailed breakdown of each feature's contribution, making it possible to compare how features influence predictions across different data points or subgroups (Michiels et al. 2024; Song et al. 2024).

However, calculating Shapley values has limitations. For models with many features, the computation becomes time-consuming because it requires evaluating all possible combinations of features—a number that grows exponentially with each additional feature. To address this, simpler estimation techniques are often used in practice. Another limitation is that Shapley values require all features to be included in the analysis, which can complicate their use in models where only a subset of features is selected. Importantly, Shapley values describe how features currently influence predictions; they do not predict how changes to feature values might alter future outcomes (Wang et al. 2024a).

To complement Shapley values, SHAP dependence plots visually illustrate how a feature's value impacts predictions. These plots display the feature's value on the horizontal axis and its corresponding SHAP value (impact on prediction) on the vertical axis. Each point represents an observation in the dataset, revealing patterns such as whether higher values of a feature consistently increase or decrease predictions. For example, in a model predicting hydrochar properties, a SHAP plot might show that higher reaction temperatures correlate with improved energy content. Such insights help identify which features matter most and guide refinements to the model or process (Bochao et al. 2023).

The SHAP library expands these capabilities by offering tools to visualize and interpret model behavior. These tools allow researchers to explore interactions between features, debug unexpected predictions, and refine models for better performance. By integrating Shapley values with intuitive visualizations, the SHAP framework bridges the gap between complex machine learning models and actionable insights, ensuring that even non-experts can understand and trust the results.

8.3 Results and Discussions

8.3.1 Model Accuracy

This chapter describes how the XGBoost algorithm (XGB) was used to develop a machine learning model for predicting six different outputs: Hydrochar yield (% w/w) (BC), Higher heating value (MJ/kg) (HHVo), Carbon content (% w/w) (Co), Oxygen content (% w/w) (Oo), Hydrogen content (% w/w) (Ho), and Ash (% w/w) (Ao). To ensure the model's reliability, tenfold cross-validation was employed during development. The prediction results, summarized in Table 8.3, show the model performed exceptionally well across all targets.

The model demonstrated strong predictive power, with test data R^2 values ranging from 0.84 to 0.97—numbers close to 1 indicate excellent accuracy. Error measurements were similarly impressive, with RMSE values between 0.03 and 0.06 and NRMSE values from 0.06 to 0.17. These low error values confirm the model's precision. Training data results were even more consistent, with perfect R^2 scores of 0.99 for all six outputs, showing the model learned the underlying patterns effectively.

Breaking down the results by output:

- Ao showed the best performance ($R^2 = 0.98$, RMSE = 0.03)
- Co followed closely ($R^2 = 0.92$, RMSE = 0.04)
- Ho and Oo performed similarly well ($R^2 = 0.90$ and 0.90 respectively)
- HHVo and BC, while slightly less accurate, still showed strong results ($R^2 = 0.88$ and 0.84)

Figure 8.3 helps visualize these results. The green lines mark where predictions fall within 10% of actual values, while red lines show the 20% range. This visualization lets us quickly assess the model's consistency and check for potential issues like

Table 8.3 Prediction accuracy in terms of R^2, RMSE, and NRMSE from tenfold cross-validation of XGB for six target outputs, from Katongtung et al. (2024c) (with permission)

Output	XGB (Train)			XGB (Test)		
	R^2	RMSE	NRMSE	R^2	RMSE	NRMSE
BC	0.9964 (0.0004)	1.0881 (−0.0638)	0.0193 (0.0012)	0.8419 (0.0495)	0.0669 (0.0095)	0.1173 (0.0165)
HHVo	0.9956 (0.0006)	0.3801 (0.0286)	0.0170 (0.0012)	0.8845 (0.0535)	0.0459 (0.0093)	0.0810 (0.0162)
Ao	0.9986 (0.0002)	0.7025 (0.0654)	0.0462 (0.0064)	0.9776 (0.0094)	0.0305 (0.0063)	0.1753 (0.0367)
Oo	0.9977 (0.0002)	0.5044 (0.0201)	0.0205 (0.0011)	0.8963 (0.0284)	0.0617 (0.0083)	0.1273 (0.0175)
Ho	0.9963 (0.0005)	0.0928 (0.0059)	0.0165 (0.0009)	0.9044 (0.0406)	0.0422 (0.0085)	0.0791 (0.0158)
Co	0.9964 (0.0006)	0.7923 (0.0657)	0.0148 (0.0013)	0.9198 (0.0315)	0.0409 (0.0102)	0.0667 (0.0175)

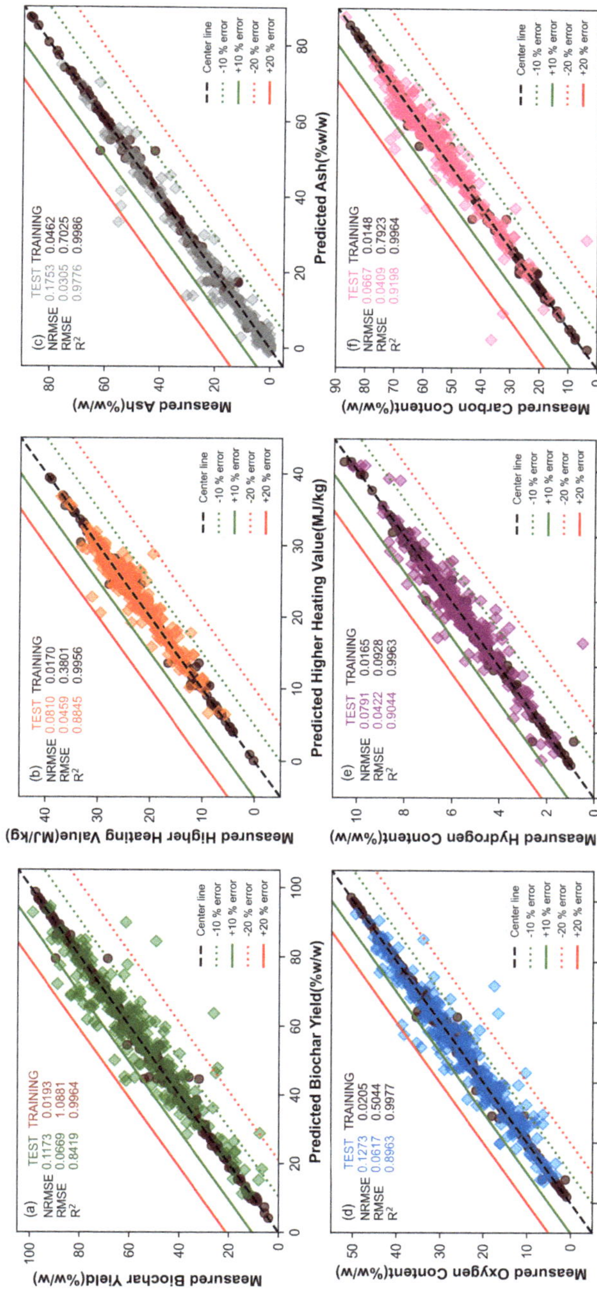

Fig. 8.3 Distribution of the training data against the test dataset: **a** BC, **b** HHVo, **c** Ao, **d** Oo, **e** Ho and **f** Co, from Katongtung et al. (2024c) (with permission)

overfitting (where a model performs well on training data but poorly on new data). The figure also compares performance between training and test datasets for each output, highlighting any significant differences in prediction quality.

8.3.2 Feature Evaluation

The impact of different factors, such as operating conditions and the properties of biomass, on hydrochar production is illustrated in Fig. 8.4. These factors are ranked by their influence, with larger changes indicating greater importance. The colors in the figure represent high (yellow) and low (blue) values for each factor.

When it comes to the yield of hydrochar (BC), the most significant factors that negatively affect output are temperature (T), nitrogen content (Ni), residence time (t), and hydrogen content (Hi). Higher temperatures speed up biomass breakdown and turn solids into gases, reducing hydrochar yield (BC) (Shafizadeh et al. 2023; Belete et al. 2021). High nitrogen content in biomass affects its composition and leads to reduced hydrochar yield because much of the nitrogen turns into soluble proteins in water (Wang et al. 2018). While longer times usually help form hydrochar through polymerization (Nakason et al. 2018), this study found that shorter times resulted in higher hydrochar yields.

For the energy content of hydrochar (HHVo), the four most important factors are temperature (T), ash content (A), initial energy content (HHVi), and carbon content (Ci), as shown in Fig. 8.4. Higher temperatures break down certain parts of biomass, which affects the energy content of the resulting hydrochar. Hydrochar produced at higher temperatures tends to have more lignin, increasing its energy content (Kang et al. 2012; Sharma et al. 2019). Biomass with higher energy and carbon content also improves the energy content of hydrochar. However, high ash content lowers the energy content due to the inorganic materials in ash (Zhu et al. 2023). Lower oxygen (Oi), carbon (Ci), and hydrogen (Hi) content in the biomass decrease ash content, enhancing combustion performance.

The oxygen content of hydrochar (Oo) is influenced by temperature (T), initial oxygen content (Oi), ash content (A), and residence time (t), as depicted in Fig. 8.4. Biomass with high oxygen content tends to produce hydrochar with higher oxygen content due to oxygen transfer during conversion. However, higher temperatures cause dehydration, reducing oxygen content (Sharma et al. 2020). Ash content acts as a catalyst, further reducing oxygen content by releasing oxygen into water (He et al. 2022a). Longer residence times enhance the removal of oxygen-rich compounds, lowering hydrochar oxygen content (Nizamuddin et al. 2017).

For the hydrogen content of hydrochar (Ho), the initial hydrogen content (Hi) has the most significant positive impact, as shown in Fig. 8.4. High hydrogen content in biomass ensures more hydrogen is retained in the hydrochar after degradation (Wang et al. 2018). Conversely, high ash content promotes dehydration reactions, reducing hydrochar hydrogen content (Gao et al. 2013). Nitrogen content input (Ni) shows a complex effect on hydrogen content output.

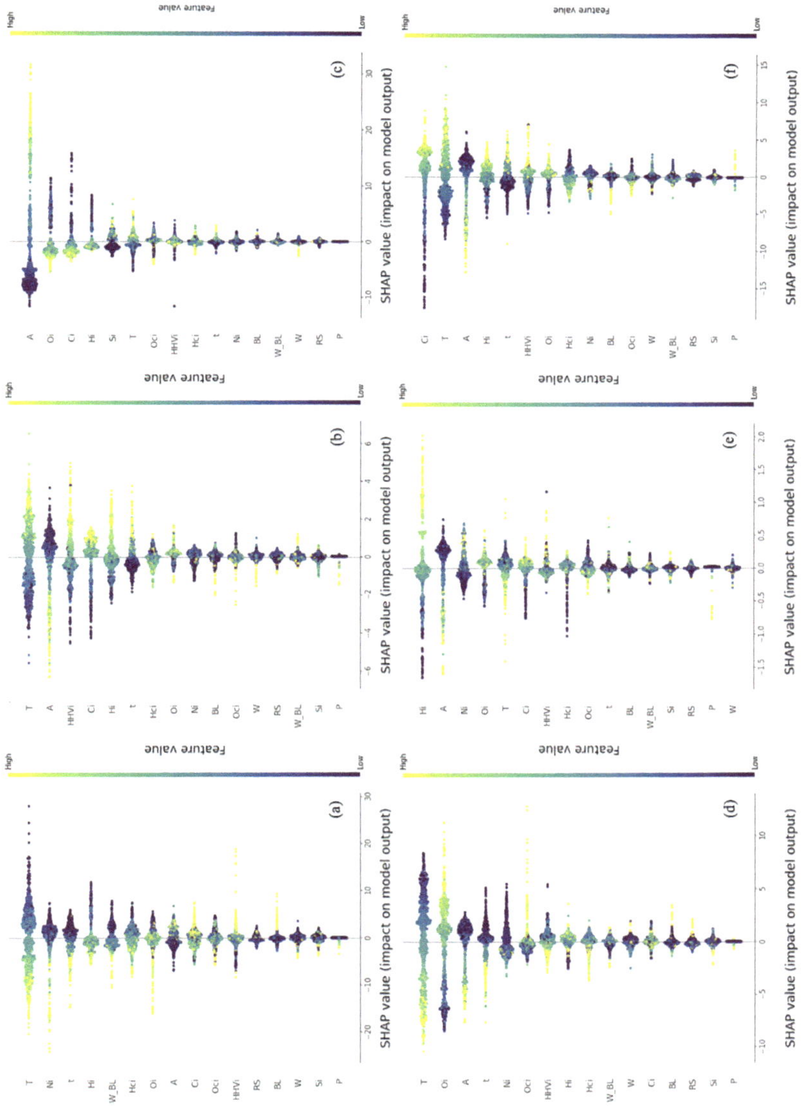

Fig. 8.4 SHAP summary plots: **a** BC, **b** HHVo, **c** Ao, **d** Oo, **e** Ho and **f** Co , from Katongtung et al. (2024c) (with permission)

Figure 8.4f highlights the factors that influence the carbon content of hydrochar. The most important factors are the carbon content of the biomass, temperature, ash content, and hydrogen content. High carbon content in biomass is crucial because it becomes the main source of carbon in hydrochar during the conversion process. Higher temperatures also increase the carbon content of hydrochar, improving its energy properties (Zhang et al. 2021). However, ash content has a negative effect on carbon content, reducing both combustion quality and energy properties (He et al. 2022b). Therefore, biomass with low ash content is ideal for producing high-quality hydrochar.

Figure 8.5 shows how different factors affect hydrochar yield and related properties. The color of the dots in these plots indicates the value of one factor interacting

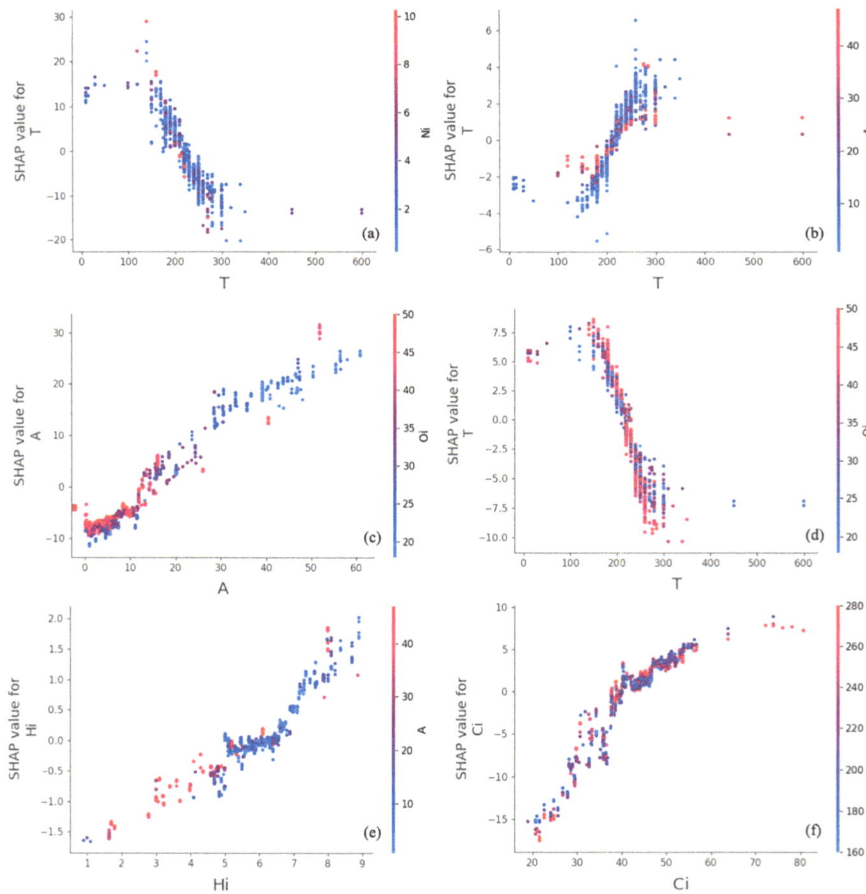

Fig. 8.5 SHAP dependence plot BC in (**a**), HHVo in (**b**), Ao in (**c**), Oo in (**d**), Ho in (**e**) and Co in (**f**) , from Katongtung et al. (2024c) (with permission)

with another, while SHAP values measure how much each input factor affects the outcome compared to a baseline scenario. Positive SHAP values mean a beneficial effect, negative values mean a harmful effect, and zero means no effect.

For hydrochar yield, SHAP values show a positive relationship with temperature below 200 °C, matching the temperatures at which certain biomass components break down (~180 °C for hemicellulose and ~200 °C for cellulose). However, the relationship with nitrogen content is more complex, peaking at a nitrogen level of 10 and a temperature of about 150 °C, as shown in Fig. 8.5a.

For the energy content of hydrochar, SHAP values increase with temperature up to about 300 °C, indicating improved energy properties. However, they decrease at higher temperatures because the high-energy components in hydrochar start to break down (Petrovic et al. 2024). The relationship between ash content and SHAP values for energy content is inconsistent. High SHAP values occur at about 300 °C when biomass ash content is low, as illustrated in Fig. 8.5b.

For ash content in hydrochar, SHAP values increase with higher ash content in the biomass, showing that most ash remains in the hydrochar. However, the relationship with oxygen content is complex, as shown in Fig. 8.5c. This suggests that while ash mostly stays in the hydrochar, its interaction with oxygen content introduces variability in the output.

For the oxygen content of hydrochar (Oo), Fig. 8.5d shows that higher temperatures above 180 °C lead to lower oxygen content. This happens because more oxygen-containing compounds are removed at higher temperatures, reducing the oxygen in the hydrochar (Jamari and Howse 2012).

Figure 8.5e highlights that high hydrogen content in the biomass and low ash content are beneficial for retaining hydrogen in hydrochar. Biomass with more hydrogen retains more hydrogen in its compounds, resulting in higher hydrogen content in the hydrochar. Low ash content also helps by minimizing dehydration reactions that turn hydrogen into water. This combination significantly increases the hydrogen content in hydrochar.

For the carbon content of hydrochar (Co), Fig. 8.5f shows that higher carbon content in the biomass leads to higher carbon content in the hydrochar. This is because carbon is retained and stabilized during the conversion process (Liu and Balasubramanian 2014). While higher temperatures generally increase carbon content, this effect is less noticeable if the biomass already has high carbon content (Liu et al. 2013). Therefore, starting with biomass rich in carbon is crucial for producing hydrochar with high carbon content.

8.4 Conclusion

This research has certain limitations. The study did not include data from mixed materials during hydrothermal carbonization (HTC). Future work could test new machine learning methods and tools to assess key factors, which might streamline development and boost accuracy. Creating easy-to-use software for HTC could also help more people apply these models in real-world settings.

This chapter showed how machine learning tools can reliably predict six critical outcomes in the HTC process. Advanced analysis methods like Shapley values revealed how different input factors—such as temperature or biomass composition—affect the results. These insights improve our understanding of HTC and offer practical strategies for improving biomass conversion methods. This could lead to more efficient energy production and reduced environmental impacts.

References

Abriha D, Srivastava PK, Szabo S (2023) Smaller is better? unduly nice accuracy assessments in roof detection using remote sensing data with machine learning and k-fold cross-validation. Heliyon 9(3). ISSN 2405-8440. https://doi.org/10.1016/j.heliyon.2023.e14045

Al Afif R, Kapidzic M, Pfeifer C (2024) Evaluation of biochar and hydrocar energy potential derived from olive mills waste: The case of montenegro. Energy 290:130234. ISSN 0360-5442. https://doi.org/10.1016/j.energy.2024.130234. https://www.sciencedirect.com/science/article/pii/S0360544224000057

Belete YZ, Mau V, Yahav Spitzer R, Posmanik R, Jassby D, Iddya A, Kassem N, Tester JW, Gross A (2021) Hydrothermal carbonization of anaerobic digestate and manure from a dairy farm on energy recovery and the fate of nutrients. Bioresou Technol 333:125164. ISSN 0960-8524. https://doi.org/10.1016/j.biortech.2021.125164. https://www.sciencedirect.com/science/article/pii/S0960852421005034

Bochao S, Cui W, Liu G, Zhou B, Zhao W (2023) A hybrid strategy of automl and shap for automated and explainable concrete strength prediction. Case Stud Constr Mater 19:e02405. ISSN 2214-5095. https://doi.org/10.1016/j.cscm.2023.e02405. https://www.sciencedirect.com/science/article/pii/S2214509523005855

Castro Garcia A, Shuo C, Cross JS (2022) Machine learning based analysis of reaction phenomena in catalytic lignin depolymerization. Bioresour Technol 345:126503. ISSN 0960-8524. https://doi.org/10.1016/j.biortech.2021.126503. https://www.sciencedirect.com/science/article/pii/S0960852421018459

Chen X, Zheng M, Cheng X, Wang C, Xu K (2022) Impact of impurities on vivianite crystallization for phosphate recovery from process water of hydrothermal carbonization of kitchen waste. Resour Conserv Recycl 185:106438. ISSN 0921-3449. https://doi.org/10.1016/j.resconrec.2022.106438. https://www.sciencedirect.com/science/article/pii/S0921344922002816

de Mora A, Diaz de Tuesta J, Pariente M, Segura Y, Puyol D, Castillo E, Lissitsyna K, Melero J, Martinez F (2024) Chemically activated hydrochars as catalysts for the treatment of htc liquor by catalytic wet air oxidation. Catalysis Today 429:114462. ISSN 0920-5861. https://doi.org/10.1016/j.cattod.2023.114462. https://www.sciencedirect.com/science/article/pii/S0920586123004868

Djandja OS, Duan P-G, Yin L-X, Wang Z-C, Duo J (2021) A novel machine learning-based approach for prediction of nitrogen content in hydrochar from hydrothermal carbonization of sewage sludge. Energy 232:121010. ISSN 0360-5442. https://doi.org/10.1016/j.energy.2021.121010. https://www.sciencedirect.com/science/article/pii/S0360544221012585

Djandja OS, Salami AA, Wang Z-C, Duo J, Yin L-X, Duan P-G (2022) Random forest-based modeling for insights on phosphorus content in hydrochar produced from hydrothermal carbonization of sewage sludge. Energy 245:123295. ISSN 0360-5442. https://doi.org/10.1016/j.energy.2022.123295. https://www.sciencedirect.com/science/article/pii/S0360544222001980

Gao Y, Wang X, Wang J, Li X, Cheng J, Yang H, Chen H (2013) Effect of residence time on chemical and structural properties of hydrochar obtained by hydrothermal carbonization of water hyacinth. Energy 58:376–383. ISSN 0360-5442. https://doi.org/10.1016/j.energy.2013.06.023. https://www.sciencedirect.com/science/article/pii/S0360544213005215

Hansted ALS, Boschert C, Hawboldt KA, Newell WJ, Yamaji FM (2024) Impact of densification process on unprocessed biomass and post-hydrothermal carbonization. Biomass Bioenergy 184:107203. ISSN 0961-9534. https://doi.org/10.1016/j.biombioe.2024.107203. https://www.sciencedirect.com/science/article/pii/S0961953424001569

He M, Cao Y, Xu Z, You S, Ruan R, Gao B, Wong K-H, Tsang DC (2022a) Process water recirculation for catalytic hydrothermal carbonization of anaerobic digestate: Water-energy-nutrient nexus. Bioresour Technol 361:127694. . ISSN 0960-8524. https://doi.org/10.1016/j.biortech.2022.127694. https://www.sciencedirect.com/science/article/pii/S0960852422010239

He M, Zhu X, Dutta S, Khanal SK, Lee KT, Masek O, Tsang DC (2022b) Catalytic co-hydrothermal carbonization of food waste digestate and yard waste for energy application and nutrient recovery. Bioresour Technol 344:126395. . ISSN 0960-8524. https://doi.org/10.1016/j.biortech.2021.126395. https://www.sciencedirect.com/science/article/pii/S0960852421017375

Jamari SS, Howse JR (2012) The effect of the hydrothermal carbonization process on palm oil empty fruit bunch. Biomass Bioenergy 47:82–90. . ISSN 0961-9534. https://doi.org/10.1016/j.biombioe.2012.09.061. https://www.sciencedirect.com/science/article/pii/S0961953412004011

Janpoom K, Suttakul P, Achariyaviriya W, Fongsamootr T, Katongtung T, Tippayawong N (2023) Investigating the influential factors in real-world energy consumption of battery electric vehicles. Energy Rep 9:316–320. . ISSN 2352-4847. https://doi.org/10.1016/j.egyr.2023.10.012. https://www.sciencedirect.com/science/article/pii/S2352484723014488. The 8th International Conference on Sustainable and Renewable Energy Engineering

Kang S, Li X, Fan J, Chang J (2012) Characterization of hydrochars produced by hydrothermal carbonization of lignin, cellulose, d-xylose, and wood meal. Ind Eng Chem Res 51(26):9023–9031. ISSN 0888-5885. https://doi.org/10.1021/ie300565d

Katongtung T, Onsree T, Tippayawong N (2022a) Machine learning prediction of biocrude yields and higher heating values from hydrothermal liquefaction of wet biomass and wastes. Bioresour Technol 344:126278. ISSN 0960-8524. https://doi.org/10.1016/j.biortech.2021.126278. https://www.sciencedirect.com/science/article/pii/S0960852421016205

Katongtung T, Phromphithak S, Onsree T, Tippayawong N, Lauterbach J (2022b) Bio-oil production from hydrothermal liquefaction of pennisetum purpureum x pennisetum typhoideum. Energy Rep 8:1170–1176. ISSN 2352-4847. https://doi.org/10.1016/j.egyr.2022.07.152. https://www.sciencedirect.com/science/article/pii/S2352484722014172. Technologies and Materials for Renewable Energy, Environment and Sustainability

Katongtung T, Onsree T, Tippayawong KY, Tippayawong N (2023) Prediction of biocrude oil yields from hydrothermal liquefaction using a gradient tree boosting machine approach with principal component analysis. Energy Rep 9:215–222. ISSN 2352-4847. https://doi.org/10.1016/j.egyr.2023.08.079. https://www.sciencedirect.com/science/article/pii/S2352484723012222. Proceedings of 2022 7th International Conference on Renewable Energy and Conservation

Katongtung T, Phromphithak S, Onsree T, Lauterbach J, Tippayawong N (2024a) Characterization of biocrude oils from hydrothermal liquefaction of de-ashed energy grass. JOM 76(1):160–170. ISSN 1543-1851. https://doi.org/10.1007/s11837-023-06234-2

Katongtung T, Phromphithak S, Onsree T, Tippayawong N (2024b) Machine learning approach for predicting hydrothermal liquefaction of lignocellulosic biomass. BioEnergy Res. ISSN 1939-1242. https://doi.org/10.1007/s12155-024-10773-0

Katongtung T, Prasertpong P, Sukpancharoen S, Sinthupinyo S, Tippayawong N (2024c) Predictive modeling for multifaceted hydrothermal carbonization of biomass. J Environ Chem Eng 12(5):114071. ISSN 2213-3437. https://doi.org/10.1016/j.jece.2024.114071. https://www.sciencedirect.com/science/article/pii/S2213343724022024

Li J, Pan L, Suvarna M, Tong YW, Wang X (2020) Fuel properties of hydrochar and pyrochar: Prediction and exploration with machine learning. Appl Energy 269:115166. ISSN 0306-2619. https://doi.org/10.1016/j.apenergy.2020.115166. https://www.sciencedirect.com/science/article/pii/S0306261920306784

Li L, Wang Y, Xu J, Flora JR, Hoque S, Berge ND (2018) Quantifying the sensitivity of feedstock properties and process conditions on hydrochar yield, carbon content, and energy content. Bioresour Technol 262:284–293. ISSN 0960-8524. https://doi.org/10.1016/j.biortech.2018.04.066. https://www.sciencedirect.com/science/article/pii/S0960852418305881

Liu C, Yue Z, Ma D, Li K, Xie Z, Zhang T, Wang J (2024a) Effect of hydrothermal carbonization on pyrolysis behavior, nutrients and metal species distribution in municipal sludge. Bioresour Technol 399:130524. ISSN 0960-8524. https://doi.org/10.1016/j.biortech.2024.130524. https://www.sciencedirect.com/science/article/pii/S096085242400227X

Liu Q, Zhang G, Yu J, Kong G, Cao T, Ji G, Zhang X, Han L (2024b) Machine learning-aided hydrothermal carbonization of biomass for coal-like hydrochar production: Parameters optimization and experimental verification. Bioresour Technol 393:130073. ISSN 0960-8524. https://doi.org/10.1016/j.biortech.2023.130073. https://www.sciencedirect.com/science/article/pii/S0960852423015018

Liu Z, Balasubramanian R (2014) Upgrading of waste biomass by hydrothermal carbonization (htc) and low temperature pyrolysis (ltp): A comparative evaluation. Appl Energy 114:857–864. ISSN 0306-2619. https://doi.org/10.1016/j.apenergy.2013.06.027. https://www.sciencedirect.com/science/article/pii/S0306261913005345

Liu Z, Quek A, Kent Hoekman S, Balasubramanian R (2013) Production of solid biochar fuel from waste biomass by hydrothermal carbonization. Fuel 103:943–949 . ISSN 0016-2361. https://doi.org/10.1016/j.fuel.2012.07.069. https://www.sciencedirect.com/science/article/pii/S0016236112006308

Mahesh T, V VK, V DK, Geman O, Margala M, Guduri M (2023) The stratified k-folds cross-validation and class-balancing methods with high-performance ensemble classifiers for breast cancer classification. Healthcare Anal 4:100247. ISSN 2772-4425. https://doi.org/10.1016/j.health.2023.100247. https://www.sciencedirect.com/science/article/pii/S2772442523001144

Mahmoud AED, Ali R, Fawzy M (2024) Insights into levofloxacin adsorption with machine learning models using nano-composite hydrochars. Chemosphere 355:141746. . ISSN 0045-6535. https://doi.org/10.1016/j.chemosphere.2024.141746. https://www.sciencedirect.com/science/article/pii/S0045653524006398

Mao F, Chen M, Zhong K, Zeng J, Liang Z (2024) An xgboost-assisted evolutionary algorithm for expensive multiobjective optimization problems. Inf Sci 666:120449. ISSN 0020-0255. https://doi.org/10.1016/j.ins.2024.120449. https://www.sciencedirect.com/science/article/pii/S0020025524003621

Michiels J, Suykens J, Vos M (2024) Explaining the model and feature dependencies by decomposition of the shapley value. Dec Supp Syst 182:114234. ISSN 0167-9236. https://doi.org/10.1016/j.dss.2024.114234. https://www.sciencedirect.com/science/article/pii/S0167923624000678

Mu L, Wang Z, Wu D, Zhao L, Yin H (2022) Prediction and evaluation of fuel properties of hydrochar from waste solid biomass: Machine learning algorithm based on proposed pso-nn model. Fuel 318:123644. ISSN 0016-2361. https://doi.org/10.1016/j.fuel.2022.123644. https://www.sciencedirect.com/science/article/pii/S0016236122005099

Nakason K, Panyapinyopol B, Kanokkantapong V, Viriya-empikul N, Kraithong W, Pavasant P (2018) Hydrothermal carbonization of unwanted biomass materials: Effect of process temperature and retention time on hydrochar and liquid fraction. J Energy Inst 91(5):786–796. ISSN 1743-9671. https://doi.org/10.1016/j.joei.2017.05.002. https://www.sciencedirect.com/science/article/pii/S174396711730185X

Nizamuddin S, Baloch HA, Griffin G, Mubarak N, Bhutto AW, Abro R, Mazari SA, Ali BS (2017) An overview of effect of process parameters on hydrothermal carbonization of biomass. Renew Sustain Energy Rev 73:1289–1299. ISSN 1364-0321. https://doi.org/10.1016/j.rser.2016.12.122. https://www.sciencedirect.com/science/article/pii/S1364032116311674

Onsree T, Tippayawong N (2020) Torrefaction of maize residue pellets with dry flue gas. BioEnergy Res 13(1):358–368 . ISSN 1939-1242. https://doi.org/10.1007/s12155-019-10058-x

Onsree T, Tippayawong N (2021) Machine learning application to predict yields of solid products from biomass torrefaction. Renew Energy 167:425–432. ISSN 0960-

1481. https://doi.org/10.1016/j.renene.2020.11.099. https://www.sciencedirect.com/science/article/pii/S0960148120318516
Peco Chacon AM, Segovia Ramirez I, Garcia Marquez FP (2023) K-nearest neighbour and k-fold cross-validation used in wind turbines for false alarm detection. Sustain Futur 6:100132. ISSN 2666-1888. https://doi.org/10.1016/j.sftr.2023.100132. https://www.sciencedirect.com/science/article/pii/S266618882300028X
Petrovic J, Ercegovic M, Simic M, Koprivica M, Dimitrijevic J, Jovanovic A, Jankovic Pantic J (2024) Hydrothermal carbonization of waste biomass: a review of hydrochar preparation and environmental application. Processes 12(1). ISSN 2227-9717. https://doi.org/10.3390/pr12010207. https://www.mdpi.com/2227-9717/12/1/207
Prasertpong P, Onsree T, Khuenkaeo N, Tippayawong N, Lauterbach J (2023) Exposing and understanding synergistic effects in co-pyrolysis of biomass and plastic waste via machine learning. Bioresour Technol 369:128419. . ISSN 0960-8524. https://doi.org/10.1016/j.biortech.2022.128419. https://www.sciencedirect.com/science/article/pii/S0960852422017527
Rasaq WA, Matyjewicz B, Swiechowski K, Lazar Z, Kupaj P, Janek T, Valentin M, Bialowiec A (2024) Food waste recycling to yarrowia biomass due to combined hydrothermal carbonization and biological treatment. J Cleaner Prod 456:142385. ISSN 0959-6526. https://doi.org/10.1016/j.jclepro.2024.142385. https://www.sciencedirect.com/science/article/pii/S095965262401833X
Reza MT, Uddin MH, Lynam JG, Coronella CJ (2014) Engineered pellets from dry torrefied and htc biochar blends. Biomass Bioenergy 63:229–238. ISSN 0961-9534. https://doi.org/10.1016/j.biombioe.2014.01.038. https://www.sciencedirect.com/science/article/pii/S0961953414000488
Shafizadeh A, Shahbeik H, Rafiee S, Moradi A, Shahbaz M, Madadi M, Li C, Peng W, Tabatabaei M, Aghbashlo M (2023) Machine learning-based characterization of hydrochar from biomass: implications for sustainable energy and material production. Fuel 347:128467. ISSN 0016-2361. https://doi.org/10.1016/j.fuel.2023.128467. https://www.sciencedirect.com/science/article/pii/S0016236123010803
Sharma HB, Panigrahi S, Dubey BK (2019) Hydrothermal carbonization of yard waste for solid bio-fuel production: study on combustion kinetic, energy properties, grindability and flowability of hydrochar. Waste Manag 91:108–119. ISSN 0956-053X. https://doi.org/10.1016/j.wasman.2019.04.056. https://www.sciencedirect.com/science/article/pii/S0956053X19302879
Sharma HB, Sarmah AK, Dubey B (2020) Hydrothermal carbonization of renewable waste biomass for solid biofuel production: A discussion on process mechanism, the influence of process parameters, environmental performance and fuel properties of hydrochar. Renew Sustain Energy Rev 123:109761. ISSN 1364-0321. https://doi.org/10.1016/j.rser.2020.109761. URL https://www.sciencedirect.com/science/article/pii/S1364032120300575
Song Z, Cao S, Yang H (2024) An interpretable framework for modeling global solar radiation using tree-based ensemble machine learning and shapley additive explanations methods. Appl Energy 364:123238. ISSN 0306-2619. https://doi.org/10.1016/j.apenergy.2024.123238. https://www.sciencedirect.com/science/article/pii/S0306261924006214
Sukpancharoen S, Katongtung T, Rattanachoung N, Tippayawong N (2023) Unlocking the potential of transesterification catalysts for biodiesel production through machine learning approach. Bioresour Technol 378:128961. ISSN 0960-8524. https://doi.org/10.1016/j.biortech.2023.128961. https://www.sciencedirect.com/science/article/pii/S0960852423003875
Taglieri L, Spera A, Gallifuoco A (2024) Green recovery of platform chemicals from hydrothermal carbonization process water. Bioresour Technol Rep 26:101815. ISSN 2589-014X. https://doi.org/10.1016/j.biteb.2024.101815. https://www.sciencedirect.com/science/article/pii/S2589014X24000562
Wang F, Ao X, Wu M, Kawata S, She J (2024a) Explainable deep learning for semg-based similar gesture recognition: a shapley-value-based solution. Inf Sci 672:120667. ISSN 0020-0255. https://doi.org/10.1016/j.ins.2024.120667. https://www.sciencedirect.com/science/article/pii/S0020025524005802

Wang K, Remon J, Jiang Z, Ding W (2024b) Recent advances in the preparation and application of biochar derived from lignocellulosic biomass: A mini review. Polymers 16(6). ISSN 2073-4360. https://doi.org/10.3390/polym16060851. https://www.mdpi.com/2073-4360/16/6/851

Wang T, Zhai Y, Zhu Y, Li C, Zeng G (2018) A review of the hydrothermal carbonization of biomass waste for hydrochar formation: process conditions, fundamentals, and physicochemical properties. Renew Sustain Energy Rev 90:223–247. ISSN 1364-0321. https://doi.org/10.1016/j.rser.2018.03.071. https://www.sciencedirect.com/science/article/pii/S1364032118301618

Wang ZH, Liu YF, Wang T, Wang JG, Liu YM, Huang QX (2024c) Intelligent prediction model of mechanical properties of ultrathin niobium strips based on xgboost ensemble learning algorithm. Comput Mater Sci 231:112579. ISSN 0927-0256. https://doi.org/10.1016/j.commatsci.2023.112579. https://www.sciencedirect.com/science/article/pii/S0927025623005736

Wilk M, Sliz M, Czerwinska K, Gajek M, Kalemba-Rec I (2024) Improvements in dewaterability and fuel properties of hydrochars derived from hydrothermal co-carbonization of sewage sludge and organic waste. Renew Energy 227:120547. ISSN 0960-1481. https://doi.org/10.1016/j.renene.2024.120547. https://www.sciencedirect.com/science/article/pii/S0960148124006128

Yeo I-K, Johnson RA (2000) A new family of power transformations to improve normality or symmetry. Biometrika 87(4):954–959. ISSN 00063444. http://www.jstor.org/stable/2673623

Zhang W, Chen Q, Chen J, Xu D, Zhan H, Peng H, Pan J, Vlaskin M, Leng L, Li H (2023) Machine learning for hydrothermal treatment of biomass: a review. Bioresour Technol 370:128547. ISSN 0960-8524. https://doi.org/10.1016/j.biortech.2022.128547. https://www.sciencedirect.com/science/article/pii/S0960852422018806

Zhang Z, Yang J, Qian J, Zhao Y, Wang T, Zhai Y (2021) Biowaste hydrothermal carbonization for hydrochar valorization: Skeleton structure, conversion pathways and clean biofuel applications. Bioresour Technol 324:124686. ISSN 0960-8524. https://doi.org/10.1016/j.biortech.2021.124686. https://www.sciencedirect.com/science/article/pii/S0960852421000249

Zhu X, Liu B, Sun L, Li R, Deng H, Zhu X, Tsang DC (2023) Machine learning-assisted exploration for carbon neutrality potential of municipal sludge recycling via hydrothermal carbonization. Bioresour Technol 369:128454. ISSN 0960-8524. https://doi.org/10.1016/j.biortech.2022.128454. https://www.sciencedirect.com/science/article/pii/S0960852422017874

Chapter 9
Hydrothermal Liquids

Abstract This chapter explores how machine learning can predict two important outcomes of hydrothermal liquefaction (HTL)—biocrude yields (BY) and energy content (measured as higher heating value, HHV)—when processing wet biomass and waste materials. The study analyzed 17 different factors that could influence these outcomes, including the biological makeup of the feedstock and the processing conditions. Researchers tested several advanced machine learning methods, using a rigorous tenfold cross-validation approach to ensure reliable results. Among these methods, the extreme gradient boosting (XGB) model proved most accurate. For biocrude yield predictions, it achieved an impressive 90% accuracy ($R^2 = 0.9$) with minimal error (NRMSE = 0.16). For energy content predictions, it maintained strong performance with 87% accuracy ($R^2 = 0.87$) and similarly low error (NRMSE = 0.04). The analysis revealed that process temperature was the single most important factor for both yield and energy predictions. However, the characteristics of the biomass feedstock itself accounted for more than half (55%) of the model's predictive power. By examining how these different factors interact, the study provides new insights that can help optimize biomass conversion processes.

Keywords Artificial intelligence · Biofuels · Hydrothermal liquefaction · Multidimensional analysis · Waste utilization

9.1 Introduction

For generations, fossil fuels have powered our world, but their limitations are now clear. As global energy needs grow, we face two critical challenges: shrinking fuel reserves and serious environmental consequences like climate change and pollution (Onsree et al. 2020; Ponnusamy et al. 2020). This urgent situation has accelerated the search for renewable alternatives, with biomass emerging as a particularly promising solution (Borugadda et al. 2020).

Scientists have developed several ways to convert biomass into usable energy. Pretreatment methods (Phromphithak et al. 2021) and torrefaction (Onsree et al. 2019; Tanyaket et al. 2020) can enhance biomass properties, while thermochemical

N. Tippayawong et al., *Machine Learning Applications for Bioenergy Conversion*, Green Energy and Technology, https://doi.org/10.1007/978-981-96-8086-3_9

processes like gasification (Elmaz et al. 2020), pyrolysis (Khuenkaeo et al. 2020; Onsree et al. 2018a), and hydrothermal processing (Saengsuriwong et al. 2021) can transform it into biofuels and biochemicals.

Hydrothermal Liquefaction: A Water-Based Solution For wet biomass materials—including food waste, fresh farm residues, and marine plants—hydrothermal liquefaction (HTL) offers special advantages. Unlike other methods, HTL works with wet materials without requiring energy-intensive drying. The process uses hot, pressurized water (180–400 °C at 5–30 MPa) to break down biomass into high-quality biocrude.

While we understand how simple materials like pure glucose behave in HTL (Kruse et al. 2007), real-world biomass presents greater complexity. Traditional mathematical models often struggle to account for all the interacting variables in these natural, variable materials.

Machine Learning: A Powerful New Tool This is where machine learning (ML) makes a difference. ML excels at handling complex, multi-factor systems and has proven valuable in biomass research:

1. Pretreatment Optimization: ML models using 23 input variables achieved 84–94% accuracy (R^2) in predicting pretreatment outcomes, showing that solvent properties account for about 20% of results (Phromphithak et al. 2021).
2. Torrefaction Predictions: Using 14 input variables, gradient boosting algorithms predicted solid yields with nearly 90% accuracy, revealing that feedstock characteristics contribute about 30% to the outcome (Onsree and Tippayawong 2020).
3. Pyrolysis and Gasification: Similar success has been shown in predicting biochar yields (84% accuracy) and gasification products like hydrogen and methane (over 85% accuracy) using biomass properties and process conditions as inputs (Pathy et al. 2020; Elmaz et al. 2020).

Researchers typically use algorithms like gradient boosting, random forests, and kernel ridge regression, working with datasets ranging from 20 to 800 samples. These tools are helping unlock the full potential of biomass as we transition to sustainable energy systems.

Advancing HTL Through Machine Learning Recent research has successfully used machine learning (ML) to improve biocrude production from hydrothermal liquefaction (HTL). Several studies demonstrate this potential:

- Gopirajan et al. (2023) used 13 key process and feedstock factors in their ML models, proving these methods work well for complex processes like HTL
- Li et al. (2021) took a broader approach, using 22 different inputs—from biomass composition to processing conditions—in their random forest model. Their system achieved 80% prediction accuracy and identified optimal conditions: 300 °C for 30 minutes, yielding 54.3% biocrude

However, these earlier studies had limitations. They focused mainly on maximizing biocrude quantity while overlooking two important aspects: predicting fuel quality (like energy content) and understanding how different factors work together to affect results.

This Chapter's Innovative Approach Our work addresses these gaps through three key advancements:

1. Dual-Purpose Modeling: We developed ML systems that predict both biocrude amounts (yield, (BY)) and energy quality (heating value, (HHV)) simultaneously
2. Comprehensive Feature Analysis: Using advanced visualization tools, we examined how each factor individually and collectively influences the HTL process with partial dependence (PDP) analysis
3. Robust Testing Framework: We evaluated four different ML algorithms (SVR, KRR, RF, and XGB) using rigorous tenfold cross-validation on datasets containing 17 different input factors

The input factors covered three main categories:

- Biological properties of the biomass
- Chemical elements present
- Processing conditions and reactor types

To understand which factors mattered most, we tested three data configurations:

- All 17 features together
- Excluding biological properties
- Excluding elemental properties

Visualizing Complex Relationships A key innovation was using partial dependence plots (PDPs) to reveal:

- How single factors affect outcomes (1D analysis)
- How factors combine to influence results (2D analysis)

This approach provides both practical optimization guidelines and fundamental scientific insights about converting wet biomass into high-quality biofuels through HTL.

9.2 Materials and Methods

Dataset and Methodology for Wet Biomass HTL Analysis This chapter's analysis builds on a comprehensive dataset of 325 experimental points from both previous work (Saengsuriwong et al. 2021) and published studies, with most measurements showing less than 5% uncertainty. The dataset focuses specifically on wet biomass materials, including:

- Aquatic biomass (macroalgae and microalgae)
- Organic wastes (food waste, animal manure)
- Municipal byproducts (sewage sludge, wastewater)

These materials differ significantly from dry lignocellulosic biomass in their biological properties, which critically influence HTL outcomes. Table 9.1 details all 17 input features analyzed, covering both feedstock characteristics and processing conditions, along with their statistical distributions.

Data Preparation and Definitions To ensure reliable machine learning results, we standardized all input features using Yeo-Johnson transformation (Yeo and Johnson 2000), as detailed in Sect. 2.7. This statistical adjustment helps ML algorithms perform better by creating more normally distributed data.

The study uses a specific definition for biocrude to maintain consistency:

Table 9.1 Set of input features/variables, from Katongtung et al. (2022) (with permission)

Variable	Name	Value	Mean	SD
Input feature				
Feedstock characteristic (on dry basis)				
Biological property				
Protein (% w/w)	Pn	0–70.7	39.3	18.2
Lipid (% w/w)	Ld	0–55.4	11.9	9.6
Carbohydrate (% w/w)	Ct	4.3–88.7	31.6	19.7
Elemental property				
Carbon content (% w/w)	C	27.9–67.1	50.5	6.7
Hydrogen content (% w/w)	H	3–10.1	7.4	1.3
Nitrogen content (% w/w)	N	1.7–12.2	7.2	2.6
Oxygen content (% w/w)	O	19.9–65.2	34.8	8.6
H-to-C atomic ratio (-)	H/C	1.1–2.3	1.7	0.2
O-to-C atomic ratio (-)	O/C	0.2–1.7	0.5	0.3
Ash (% w/w)	Ash	2–47.5	15.9	11.8
Operating condition				
Residence time (min)	Time	1.4–120	36	23.7
Temperature (°C)	T	150–400	315.4	45.2
(Initial) pressure (MPa)	P	0–10	1.4	2.1
Reactor size (mL)	RS	2.7–1800	188	374.2
Biomass loading (g)	BL	0.1–25	6.6	6.5
Water (mL)	W	1–475	60.3	96.9
Solvent-to-BL ratio (-)	S/BL	2.6–100	11.4	15.9
Output target				
Biocrude yield (% w/w)	BY	3–69.2	36.2	15.6
Higher heating value (MJ/kg)	HHV	25.4–46.8	30.9	3.8

- Biocrude: The organic fraction separated from water-soluble components using solvents like acetone or dichloromethane
- Biocrude yield (y_{BY}) was calculated using the following method:

$$y_{BY} = \frac{m_{BY}}{M} \tag{9.1}$$

where m_{BY} is the mass of biocrudes and M is the initial mass of feedstocks on a dry basis. Similarly, the HHV of the biocrude was recalculated as follows (Rajagopal et al. 2021):

$$HHV = 0.338\text{C} + 1.428\,(\text{H} - \text{O}/8) \tag{9.2}$$

where, C, H, and O are the carbon, hydrogen, and oxygen contents of the biocrude (in w/w).

9.2.1 ML Model Development

This chapter employed four advanced machine learning algorithms to analyze hydrothermal liquefaction data:

Algorithms Used:

- Support Vector Regression (SVR)
- Kernel Ridge Regression (KRR)
- Random Forest (RF)
- XGBoost (XGB)

The models were built using Python's Scikit-learn library (Pedregosa et al. 2011) with XGBoost implemented through its dedicated open-source package (Chen and Guestrin 2016). All computations ran on an AMD Ryzen 5 processor operating at 2.10 GHz.

Rigorous Validation Approach We implemented a robust tenfold cross-validation method to thoroughly evaluate model performance:

(1) The complete dataset was randomly shuffled and divided into 10 equal groups (folds)

(2) For each validation cycle:

- Models trained on 9 folds (90% of data)
- Tested on the remaining fold (10% of data)

(3) This process was repeated 10 times, with each fold serving as the test set once

(4) Final performance metrics represented averages across all iterations

This approach ensures:

- Every data point contributes to both training and testing
- Unbiased assessment of prediction accuracy
- Reliable representation of the dataset's statistical properties (Elmaz et al. 2020)

Performance Evaluation Metrics Model accuracy was assessed using three key metrics:

1. R^2 (Coefficient of Determination): Measures how well the model explains variance in the data (values closer to 1 indicate better fit)
2. RMSE (Root Mean Square Error): Quantifies prediction errors in original units
3. NRMSE (Normalized RMSE): Calculated as RMSE divided by the mean observed value, $NRMSE = RMSE/\bar{y}$ where $\bar{y}$ represents the mean of the observed values, allowing comparison across different scales (Onsree and Tippayawong 2020)

This comprehensive validation framework provides reliable insights into each algorithm's predictive capabilities for hydrothermal liquefaction outcomes.

9.3 Selected ML Algorithm

The four machine learning algorithms used in this study each have unique mathematical approaches that make them particularly suited for different aspects of data analysis:

1. Support Vector Regression (SVR): Effective for high-dimensional data, using kernel tricks to model complex relationships
2. Kernel Ridge Regression (KRR): Combines ridge regression with kernel methods for handling nonlinear patterns
3. Random Forest (RF): An ensemble method that builds multiple decision trees for robust predictions
4. XGBoost (XGB): A powerful gradient-boosted trees algorithm known for its high performance

(For detailed mathematical foundations of these algorithms, refer to Chap. 2)

Hyperparameter Optimization Process To maximize each algorithm's predictive accuracy, we implemented a rigorous optimization procedure:

- Full-factorial grid search: Systematically tested all possible combinations of hyperparameter values
- Nested tenfold cross-validation:
  - Outer loop: Evaluated model performance
  - Inner loop: Tuned hyperparameters
  - This prevented overfitting (Pedregosa et al. 2011)

Table 9.2 Optimized hyperparameters of the four algorithms to predict BY and HHV from Katongtung et al. (2022) (with permission)

ML Algorithm	Hyperparameters	Optimized value		Execution Time[a]
		BY	HHV	
SVR	C	1	1	~30
	ϵ	9×10^{-5}	9×10^{-5}	
	Kernel	"RBF"	"RBF"	
	"tol"	9×10^{-6}	5×10^{-6}	
	"random_state"	66	68	
KRR	α	9.6×10^{-4}	9.5×10^{-4}	~20
	Kernel	"RBF"	"RBF"	
	γ	5.1×10^{-3}	5.1×10^{-3}	
	"random_state"	66	68	
RFR	"n_estimators"	149	149	~300
	"max_depth"	85	85	
	"max_leaf_nodes"	480	480	
	"min_samples_split"	7	5	
	"random_state"	66	68	
XGB	"learning_rate"	7.0×10^{-2}	7.9×10^{-2}	~180
	"n_estimators"	270	270	
	γ	3×10^{-5}	3×10^{-5}	
	"max_depth"	80	80	
	"min_child_weight"	13	13	
	"subsample"	0.66	0.66	
	"colsample_bytree"	0.8	0.8	
	"colsample_bylevel"	0.9	0.9	
	"colsample_bynode"	0.6	0.6	
	"random_state"	66	68	

Note other unlisted hyperparameters were kept at default values.
[a]The time (in ms) is spent in each algorithm to process both training and testing tasks in tenfold cross-validation.

The resulting optimal hyperparameters and computational requirements for each algorithm are documented in Table 9.2. This optimization process ensured that each model achieved its best possible performance while maintaining generalization capability.

9.3.1 Feature Evaluation and Interpretation

To understand how different biomass characteristics and process conditions affect hydrothermal liquefaction (HTL) outcomes, we analyzed the machine learning models using three carefully designed datasets:

1. Complete Feature Set (Case 1): All 17 original features including both biological properties and elemental compositions
2. Non-Biological Features (Case 2): All features excluding biological characteristics
3. Non-Elemental Features (Case 3): All features excluding elemental compositions

Model Training and Feature Analysis For each dataset configuration, we performed:

- Comprehensive hyperparameter optimization using grid search
- Rigorous tenfold cross-validation to ensure reliable results
- Feature importance analysis using each algorithm's built-in methods (Igual and Segui 2024; Phromphithak et al. 2021)

This process identified nine key features that most significantly influenced the predictions of both biocrude yield (BY) and energy content (HHV). Understanding these influential factors provides valuable insights into what drives HTL performance.

Visualizing Feature Relationships Advanced visualization techniques were employed to better understand these relationships:

- One-dimensional PDPs: Show how changing individual features affects HTL outcomes
- Two-dimensional PDPs: Reveal complex interactions between pairs of features (Zhu et al. 2019)

For example, these plots might show:

- How temperature and reaction time work together to influence yields
- The relationship between biomass carbon content and energy output
- Threshold effects where small changes create significant impacts

This multifaceted analysis approach allows us to move beyond simple predictions and truly understand the underlying factors controlling HTL performance. The findings provide practical guidance for optimizing biomass conversion processes while advancing fundamental knowledge of hydrothermal liquefaction systems.

9.4 Results and Discussion

9.4.1 Model Selection and Accuracy

Table 9.3 shows how well different models (XGB, RF, KRR, and SVR) predict results when using their best settings across three different sets of data. The results highlight significant changes in prediction accuracy when comparing the full set of 17 features (Case 1) to sets where certain features are left out (Cases 2 and 3).

Table 9.3 Prediction accuracy in terms of R^2, RMSE, and NRMSE from tenfold cross-validation of XGB, RF, KRR, and SVR algorithms for all three cases of input features, from Katongtung et al. (2022) (with permission)

Model	Input	BY			HHV		
		R^2(SD.)	RMSE (SD.)	NRMSE (SD.)	R^2(SD.)	RMSE (SD.)	NRMSE (SD.)
XGB	Case 1	0.8711 (−0.0446)	5.4137 (−1.3129)	0.1772 (−0.0354)	0.8194 (−0.0485)	1.9846 (−0.8847)	0.0559 (−0.0263)
	Case 2	0.8416 (−0.0466)	5.9923 (−1.1112)	0.1976 (−0.0319)	0.8187 (−0.0484)	1.9994 (−0.9112)	0.0563 (−0.027)
	Case 3	0.8666 (−0.0407)	5.5303 (−1.2229)	0.1812 (−0.0316)	0.7982 (−0.0811)	2.1092 (−1.0902)	0.0595 (−0.0323)
RF	Case 1	0.8425 (−0.0487)	6.0073 (−1.3018)	0.1967 (−0.0323)	0.7353 (−0.121)	2.2857 (−0.7445)	0.0642 (−0.0222)
	Case 2	0.8196 (−0.0577)	6.4225 (−1.3588)	0.2109 (−0.0368)	0.7319 (−0.1282)	2.297 (−0.7601)	0.0645 (−0.0227)
	Case 3	0.8412 (−0.0524)	6.0252 (−1.3801)	0.1969 (−0.0322)	0.7954 (−0.0292)	2.0919 (−0.7659)	0.0588 (−0.0228)
KRR	Case 1	0.8201 (−0.0878)	6.3053 (−1.6961)	0.2064 (−0.0453)	0.6567 (−0.2563)	2.5142 (−0.8931)	0.0705 (−0.0256)
	Case 2	0.7644 (−0.1051)	7.2417 (−1.6643)	0.2378 (−0.044)	0.5742 (−0.2843)	2.8384 (−1.0404)	0.0797 (−0.0303)
	Case 3	0.729 (−0.0922)	7.7902 (−1.4068)	0.2574 (−0.0437)	0.5183 (−0.119)	3.2136 (−1.3033)	0.0904 (−0.0387)
SVR	Case 1	0.8457 (−0.0474)	5.9615 (−1.3387)	0.1949 (−0.0319)	0.7134 (−0.1413)	2.4259 (−0.9328)	0.0682 (−0.0272)
	Case 2	0.7096 (−0.1334)	8.0604 (−2.2868)	0.2636 (−0.0659)	0.5352 (−0.2518)	3.0169 (−1.0591)	0.0847 (-0.0307)
	Case 3	0.6722 (−0.1565)	8.561 (−2.3757)	0.2791 (−0.0643)	0.4201 (−0.189)	3.3705 (−0.915)	0.0946 (−0.0274)

For instance, using the Random Forest model, the accuracy measure (R^2) for predicting biocrude yield (BY) dropped from 0.84 in Case 1 to 0.82 in Case 2. However, for predicting the higher heating value (HHV), the accuracy improved from 0.74 in Case 1 to 0.80 in Case 3. The error measure (RMSE) for BY increased from 6.01 in Case 1 to 6.42 in Case 3, while the error for HHV slightly decreased from 2.28 in Case 1 to 2.09 in Case 3.

Similarly, for the KRR and SVR models, the accuracy for BY was between 0.82 and 0.85 in Case 1, but fell to between 0.67 and 0.73 in Case 3. For HHV, the accuracy initially ranged from 0.66 to 0.71 in Case 1 but dropped to between 0.54 and 0.57 in Case 3. The error for BY increased from 5.96 to 6.31 in Case 1 to 7.79–8.56 in Case 3, while for HHV, it rose from 2.43 3to 2.51 in Case 1 to 3.21–3.37 in Case 3.

Among all models, XGB performed the best overall based on repeated testing. For BY, the accuracy ranged from 0.84 to 0.87, and the error was between 5.41 and 5.99. For HHV, the accuracy was around 0.80 to 0.82, and the error ranged from 1.98

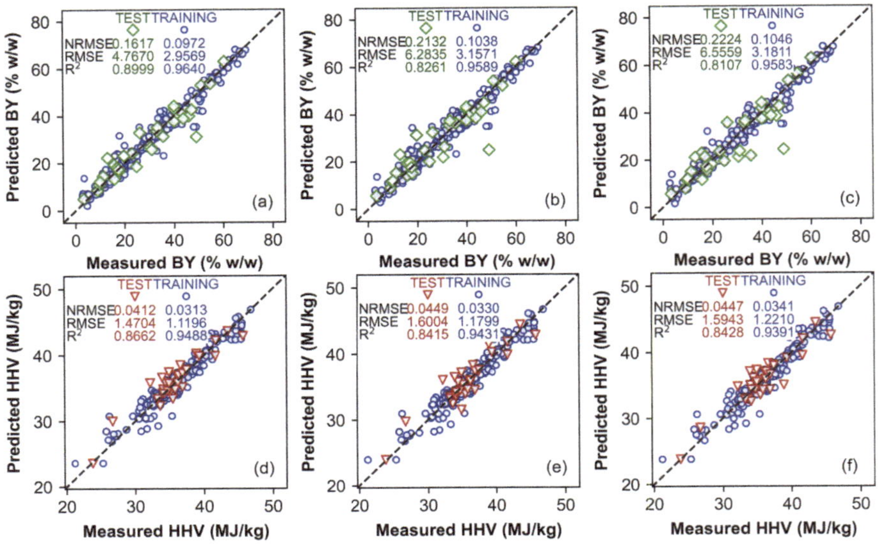

Fig. 9.1 Distribution of the prediction data against the test dataset from the first fold in ten folds of cross-validation of BY in (**a**), (**c**), and (**e**), as well as HHV in (**b**), (**d**), and (**f**) for Case 1 in (**a**) and (**b**), Case 2 in (**c**) and (**d**) and Case 3 in (**e**) and (**f**). The black trend line represents the positions of the predicted values are exactly equal to the test values, from Katongtung et al. (2022) (with permission)

to 2.11. These results show that XGB is particularly good at predicting both BY and HHV of biocrudes.

Figure 9.1 compares the predicted and actual values for biocrude yield (BY) and higher heating value (HHV) using the XGB model under both training and testing conditions. The test data used here comes from the first part of a method called tenfold cross-validation, applied to all three sets of input features. It's clear that the model performed better on the training data than on the test data across all cases.

During training, the model was very accurate for BY predictions, with an R^2 ranging from 0.96 to 0.97, and a normalized root mean square error (NRMSE) between 0.10 and 0.11. In testing, the accuracy was slightly lower, with R^2 values from 0.81 to 0.90 and an NRMSE of 0.16 to 0.23. For HHV predictions, the training data showed R^2 values of 0.94 to 0.95 and an NRMSE of about 0.03. On the test data, R^2 ranged from 0.84 to 0.87, with an NRMSE around 0.04.

When comparing different input datasets, the XGB model worked best when using all 17 features (Case 1). For BY, Case 1 had an R^2 near 0.90 and an NRMSE of 0.16. Cases 2 and 3 had similar results, with R^2 values around 0.82 and an average NRMSE of 0.22. For HHV, Case 1 provided the best results with an R^2 of 0.87 and an NRMSE of 0.04. Cases 2 and 3 had slightly lower R^2 values of about 0.84 and slightly higher NRMSE values compared to Case 1. These results suggest that including more relevant features in the input dataset enhances model prediction accuracy (Onsree and Tippayawong 2020; Phromphithak et al. 2021).

Overall, the XGB algorithm is effective in predicting the yield and HHV of biocrudes from hydrothermal liquefaction of wet biomass. Its accuracy is comparable to models developed for specific biomass types or other machine learning models used in different biomass conversion processes (Gopirajan et al. 2023; Hong et al. 2021; Zhu et al. 2019). This highlights the versatility and reliability of XGB in modeling complex processes involving diverse feedstock properties and operating conditions.

9.4.2 Feature Evaluation

The importance of features in the XGB model was analyzed using a method that plots feature importance, with the results shown in Fig. 9.2. For both biocrude yield (BY) and higher heating value (HHV), the top five input features were identified as the main contributors to the model's predictions. Among these, the temperature (T) feature was ranked first, contributing about 10% to the total. Following temperature, features like Pn, Ld, and Ct—related to biological properties—ranked highly, collectively contributing about 25%. The H/C ratio, representing elemental properties, secured the fifth position.

In total, operating conditions contributed about 45% to BY predictions and approximately 40% to HHV predictions. Notably, features related to elemental properties and biological properties had a nearly equal influence on both BY and HHV predictions. This balanced contribution aligns with the results shown in Table 9.3 and Fig. 9.1, highlighting the essential role of these feature categories. Without either elemental or biological properties, the models showed a similar reduction in prediction accuracy.

Temperature plays a crucial role in hydrothermal liquefaction (HTL) reactions (Chen et al. 2014). Higher temperatures enhance biocrude yield through reactions

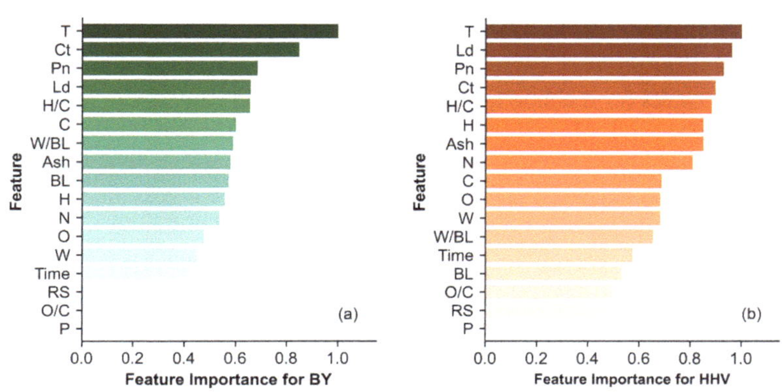

Fig. 9.2 Feature importance of **a** BY and **b** HHV, from Katongtung et al. (2022) (with permission)

that break down biocrude into lighter gases and produce biochar. The influence of temperature on BY and HHV is closely tied to the composition of the feedstock, which typically contains a mix of proteins, lipids, and carbohydrates (Biswas et al. 2017; Reddy et al. 2016).

The chemical reactions during wet biomass HTL are complex. Feedstock composition and the degree of component separation significantly impact reaction mechanisms and product profiles. Proteins, a major component of wet biomass, contribute significantly to the nitrogen content in microalgae and influence the characteristics of biocrude.

At lower HTL temperatures, microbial cells degrade more readily (Jazrawi et al. 2015). Biological macro-molecules hydrolyze into simpler compounds (Yang et al. 2018). As temperatures increase, these compounds can undergo further reactions, forming new compounds that contribute to BY and HHV. High temperatures promote these reactions, which are important for understanding how biocrude properties are affected during HTL (Tang et al. 2020).

Carbon (C) and hydrogen (H) are two of the most important elements influencing the XGB model's predictions. Carbon comes from atmospheric CO_2, which plants absorb during photosynthesis. It plays a key role in determining the energy content of biomass. When biomass burns, most of its carbon is released back into the atmosphere as CO_2, completing the natural carbon cycle.

Hydrogen is another essential component of biomass. It is found in the chemical structure of carbohydrates and phenolic polymers that make up plant material. During combustion, hydrogen turns into water (H_2O) and significantly contributes to the energy output. These elemental properties highlight how biomass composition affects its energy potential and efficiency in energy conversion.

9.4.3 PDPs and Interpretation

To better understand how the most important factors influence the predictions of biocrude yield (BY) and higher heating value (HHV), one-dimensional (1D) and two-dimensional (2D) partial dependence plots (PDPs) were used. Figure 9.3 shows the 1D PDP results for the nine key features affecting BY and HHV.

In these plots, the blue tick marks along the x-axis represent the density of data points for each feature. These markers provide important context, helping to interpret the trends shown in the PDPs. Without them, the observed patterns might not clearly reflect the dataset. These visual tools help illustrate how individual factors contribute to the model's predictions, offering valuable insights into their role in determining BY and HHV.

For biocrude yield (BY), as shown in Fig. 9.3a, temperature (T) plays a significant role. The impact of temperature on BY increases sharply between 245 °C and 300 °C, but beyond 300 °C, the effect becomes less pronounced. This suggests that within the 245–300 °C range, temperature has a stronger influence on hydrothermal liquefaction (HTL) reactions compared to higher temperatures. As temperature

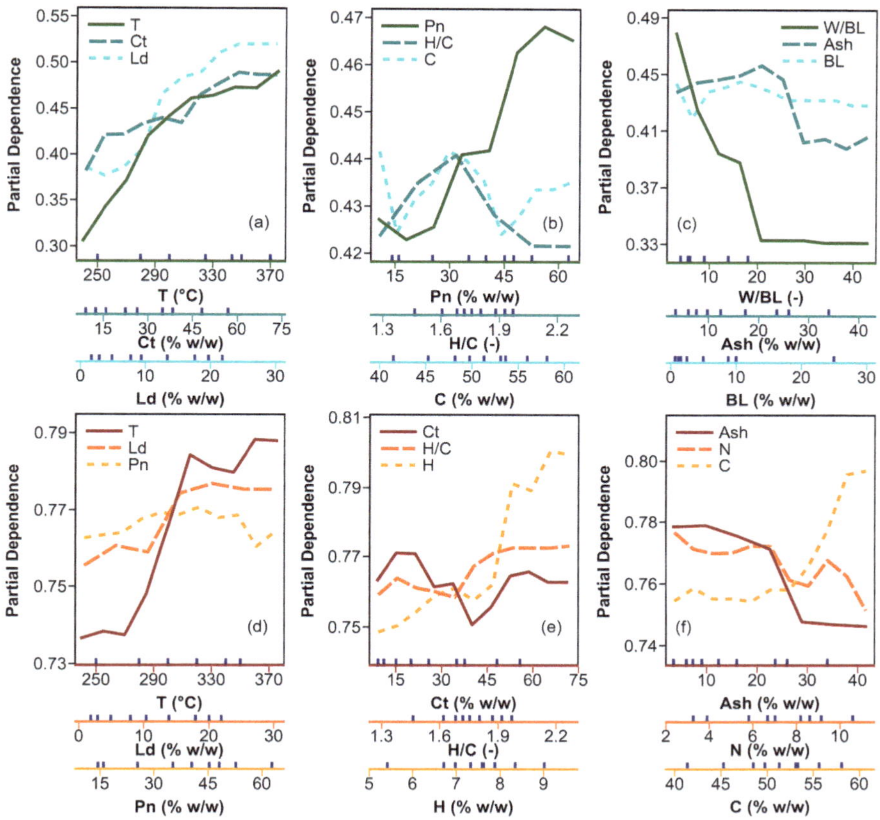

Fig. 9.3 One-dimensional PDPs of the nine most important features on predictions of BY in (**a**), (**b**), and (**c**), as well as of HHV in (**d**), (**e**), and (**f**), by XGB, from Katongtung et al. (2022) (with permission)

rises, biomass undergoes more degradation, leading to increased conversion into biocrudes and water-soluble products (Mishra and Mohanty 2020). This is due to enhanced hydrolysis and decomposition reactions that break biomass into smaller compounds, forming biocrudes. At higher temperatures, the energy needed to break chemical bonds is more easily overcome, leading to significant depolymerization and increased biocrude yields (Mehrabadi et al. 2017).

Figure 9.3a and b show that carbohydrate content (Ct), protein content (Pn), and lipid content (Ld) all positively influence BY. The relationship between Ct and BY is nearly linear. While carbohydrates mainly contribute to residual solids and soluble organic matter during HTL, their interaction with lipids enhances biocrude production. As a result, higher carbohydrate content indirectly boosts BY. The effect of Pn initially declines slightly but then increases sharply once lipid content exceeds 20% w/w, stabilizing beyond 50% w/w. The degradation of proteins during

HTL enhances biocrude yields, particularly due to the hydrophobic components of protein-carbohydrate materials, which increase nitrogen (N) content in the biocrude (Aierzhati et al. 2019; Torri et al. 2012). The Ld feature follows a similar trend, first decreasing slightly, then rising after lipid content surpasses 2.5% w/w, and leveling off beyond 25% w/w. Research shows that lipids are a primary source of biocrude, so higher lipid content correlates with greater biocrude production (Chen et al. 2014).

Figure 9.3b and c illustrate how the hydrogen-to-carbon ratio (H/C), carbon content (C), and ash content affect BY. The H/C ratio has a positive impact on BY up to a value of 1.75, after which its effect turns negative. The influence of H/C and C depends on the reactions occurring during HTL. The HTL process concentrates C and H in the biocrude while removing oxygen (O) and nitrogen (N), leading to various reactions such as deoxygenation, denitrogenation, and dehydration (Tian et al. 2015). Ash content, on the other hand, generally reduces BY. At low levels (<20% w/w), ash may act as a catalyst, enhancing HTL reactions and increasing biocrude production. However, when ash exceeds 20% w/w, it has a negative effect on BY.

Figure 9.3c also shows the effects of the water-to-biomass ratio (W/BL) and biomass loading (BL). Both features negatively impact BY as their values increase. A higher W/BL means more water in the HTL process, which can reduce biocrude production due to competing hydrolysis and repolymerization reactions (Neveux et al. 2014). Additionally, excessive water usage raises processing costs. The effect of BL fluctuates around 2.5 g before gradually declining. A higher BL increases gas production but can limit heat and mass transfer, reducing the extent of hydrothermal degradation and ultimately leading to lower biocrude yields.

In Fig. 9.3d, the effect of temperature (T) on the higher heating value (HHV) of biocrudes follows a similar pattern to its influence on biocrude yield (BY). HHV increases steadily after 270 °C and stabilizes around 300 °C. This rise in HHV is linked to the deoxygenation process, which removes oxygen from biomass and improves fuel quality at higher temperatures (Saengsuriwong et al. 2021).

Among the biological properties, lipid content (Ld) consistently enhances HHV, meaning that higher lipid levels improve the energy quality of biocrudes produced from wet biomass during hydrothermal liquefaction (HTL) (Yang et al. 2019). In contrast, protein content (Pn) and carbohydrate content (Ct) show more complex patterns. The Pn feature positively affects HHV within the 15%-30% w/w range and again above 30% w/w. While proteins can boost biocrude yield, biocrudes derived from high-protein feedstocks may require additional refining due to their complex chemical composition.

Figure 9.3e shows that Ct positively influences HHV when its values are below 20% w/w and between 40% and 60% w/w. However, between 20% and 40% w/w, its effect declines before stabilizing beyond 60% w/w. Lipids, proteins, and carbohydrates all contribute to hydrogenation reactions, which enhance HHV. Overall, the presence of either Pn or Ct in the HTL process benefits hydrogenation, improving biocrude quality.

Regarding elemental properties, hydrogen-to-carbon ratio (H/C), hydrogen (H), and carbon (C) all positively impact HHV, as shown in Fig. 9.4c and f. On the other hand, ash and nitrogen (N) have a negative effect (Fig. 9.3f). Studies indicate that

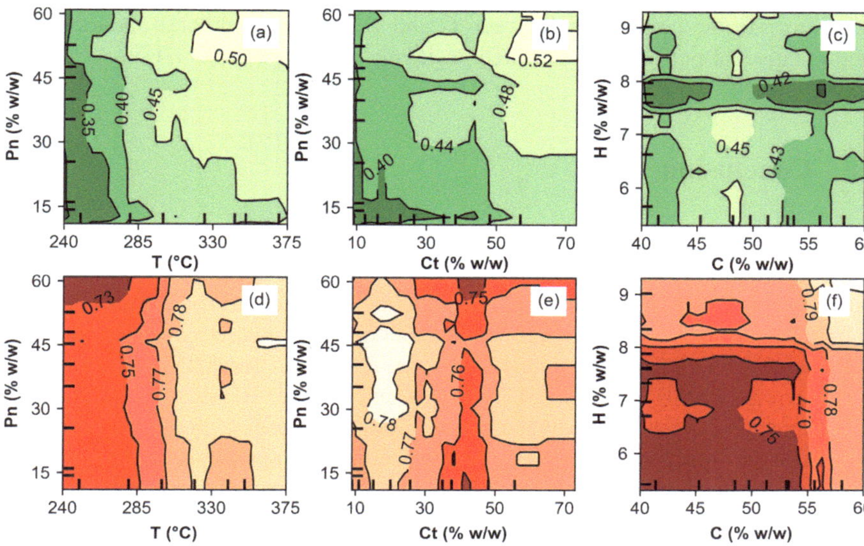

Fig. 9.4 Two-dimensional PDPs of the XGB predictions for BY in (**a**), (**b**), and (**c**), as well as HHV in (**d**), (**e**), and (**f**), from Katongtung et al. (2022) (with permission)

higher levels of C and H, along with reduced oxygen content, increase the energy density of biocrudes. A high H value generally improves fuel quality, as fuels with a higher H/C ratio tend to perform better.

Hydrogen in HTL reactions primarily comes from the water used in the process. Nitrogen content in wet biomass gradually decreases, but some nitrogen from proteins can remain in the biocrude. Peptide bonds in proteins are more stable than the glycosidic bonds found in carbohydrates, making them less prone to hydrolysis. However, at high temperatures, nitrogenous fragments can convert into biocrude through reactions like the Maillard reaction, which involves interactions between sugars and amino acids, producing nitrogen-containing compounds in limited amounts.

These trends are consistent with findings by Garcia Alba et al. (2012), who observed that nitrogen increases while oxygen decreases in biocrudes, affecting their HHV. Additionally, HHV declines as ash content rises, since ash does not contribute to calorific value. Even though only a small amount of ash transfers into biocrudes during HTL, excessive ash in the feedstock lowers the overall HHV (Chen et al. 2017).

To examine how two features interact in relation to BY and HHV, two-dimensional partial dependence plots (PDPs) were used. Figure 9.4a and d illustrate the interaction between T and Pn, showing that higher temperatures promote protein breakdown into amino acids, influencing both yield and HHV.

Figure 9.3e highlights the relationship between Ct and Pn. As noted by Torri et al. (2012), interactions between proteins and carbohydrates can enhance both BY and HHV at higher temperatures due to Maillard reactions. Finally, Fig. 9.4c and f demonstrate the connection between C and H. The HTL process facilitates the accumulation of both elements, which positively impacts BY and HHV.

9.5 Conclusion

The XGB algorithm was successfully applied to predict biocrude yield (BY) and a higher heating value (HHV) from the hydrothermal liquefaction (HTL) of wet biomass and waste materials. The model used input features based on feedstock characteristics and operating conditions, achieving high predictive accuracy. It produced an R^2 value of approximately 0.90 for BY and 0.87 for HHV, with root mean square errors (RMSE) of around 4.77 for BY and 1.47 for HHV.

Among all the factors analyzed, temperature emerged as the most influential variable in the model's predictions. However, feedstock properties collectively played a greater role in determining outcomes. Understanding how each feature contributes to BY and HHV, as well as their interactions, provides valuable insights into the HTL process.

These findings demonstrate the impact of different feedstock compositions and process conditions on biocrude production, offering a solid foundation for optimizing HTL operations to maximize energy recovery.

References

Aierzhati A, Stablein MJ, Wu NE, Kuo C-T, Si B, Kang X, Zhang Y (2019) Experimental and model enhancement of food waste hydrothermal liquefaction with combined effects of biochemical composition and reaction conditions. Bioresour Technol 284:139–147. ISSN 0960-8524. https://doi.org/10.1016/j.biortech.2019.03.076. https://www.sciencedirect.com/science/article/pii/S0960852419304353

Biswas B, Arun Kumar A, Bisht Y, Singh R, Kumar J, Bhaskar T (2017) Effects of temperature and solvent on hydrothermal liquefaction of sargassum tenerrimum algae. Bioresour Technol 242:344–350. ISSN 0960-8524. https://doi.org/10.1016/j.biortech.2017.03.045. https://www.sciencedirect.com/science/article/pii/S0960852417303152. Special Issue on International Conference on Current Trends in Biotechnology 'I&' post ICCB-2016 conference on Strategies for Environmental Protection and Management (ICSEPM-2016)

Borugadda VB, Chand R, Dalai AK (2020) Screening suitable refinery distillates for blending with htl bio-crude and evaluating the co-processing potential at petroleum refineries. Energy Conver Manag 222:113186. ISSN 0196-8904. https://doi.org/10.1016/j.enconman.2020.113186. https://www.sciencedirect.com/science/article/pii/S0196890420307305

Chen T, Guestrin C (2016) Xgboost: a scalable tree boosting system. In: Proceedings of the 22nd ACM SIGKDD international conference on knowledge discovery and data mining, KDD '16, pp 785 – 794, New York, NY, USA, 2016. Association for Computing Machinery. ISBN 9781450342322. https://doi.org/10.1145/2939672.2939785. https://doi.org/10.1145/2939672.2939785

Chen W-T, Zhang Y, Zhang J, Yu G, Schideman LC, Zhang P, Minarick M (2014) Hydrothermal liquefaction of mixed-culture algal biomass from wastewater treatment system into bio-crude oil. Bioresour Technol 152:130–139. ISSN 0960-8524. https://doi.org/10.1016/j.biortech.2013.10.111. https://www.sciencedirect.com/science/article/pii/S0960852413016982

Chen W-T, Qian W, Zhang Y, Mazur Z, Kuo C-T, Scheppe K, Schideman LC, Sharma BK (2017) Effect of ash on hydrothermal liquefaction of high-ash content algal biomass. Algal Res 25:297–306. ISSN 2211-9264. https://doi.org/10.1016/j.algal.2017.05.010. https://www.sciencedirect.com/science/article/pii/S2211926416308062

Cheng J, Huang R, Yu T, Li T, Zhou J, Cen K (2014) Biodiesel production from lipids in wet microalgae with microwave irradiation and bio-crude production from algal residue through hydrothermal liquefaction. Bioresour Technol 151:415–418. ISSN 0960-8524. https://doi.org/10.1016/j.biortech.2013.10.033. https://www.sciencedirect.com/science/article/pii/S0960852413016131

Elmaz F, Yucel O, Mutlu AY (2020) Predictive modeling of biomass gasification with machine learning-based regression methods. Energy 191:116541. ISSN 0360-5442. https://doi.org/10.1016/j.energy.2019.116541. https://www.sciencedirect.com/science/article/pii/S0360544219322364

Garcia Alba L, Torri C, Samori C, van der Spek J, Fabbri D, Kersten SRA, Brilman DWFW (2012) Hydrothermal treatment (htt) of microalgae: evaluation of the process as conversion method in an algae biorefinery concept. Energy Fuels 26(1):642–657. ISSN 0887-0624. https://doi.org/10.1021/ef201415s. https://doi.org/10.1021/ef201415s

Garcia Nieto P, Garcia-Gonzalo E, Sanchez Lasheras F, Paredes-Sanchez J, Riesgo Fernandez P (2019) Forecast of the higher heating value in biomass torrefaction by means of machine learning techniques. J Comput Appl Math 357:284–301. ISSN 0377-0427. https://doi.org/10.1016/j.cam.2019.03.009. https://www.sciencedirect.com/science/article/pii/S0377042719301281

Gollakota A, Kishore N, Gu S (2018) A review on hydrothermal liquefaction of biomass. Renew Sustain Energy Rev 81:1378–1392. ISSN 1364-0321. https://doi.org/10.1016/j.rser.2017.05.178. https://www.sciencedirect.com/science/article/pii/S1364032117308146

Gopirajan PV, Gopinath KP, Sivaranjani G, Arun J (2023) Optimization of hydrothermal liquefaction process through machine learning approach: process conditions and oil yield. Biomass Conversion and Biorefinery 13(2):1213–1222. ISSN 2190-6823. https://doi.org/10.1007/s13399-020-01233-8. https://doi.org/10.1007/s13399-020-01233-8

Han L, Lin Q, Liu G, Han D, Niu L, Su D (2019) Lipids promote glycated phospholipid formation by inducing hydroxyl radicals in a maillard reaction model system. J Agric Food Chem 67(28):7961–7967. ISSN 0021-8561. https://doi.org/10.1021/acs.jafc.9b02771. https://doi.org/10.1021/acs.jafc.9b02771

Hong C, Wang Z, Si Y, Li Z, Xing Y, Hu J, Li Y (2021) Preparation of bio-oils by hydrothermal liquefaction (htl) of penicillin fermentation residue (pr): Optimization of conditions and mechanistic studies. Sci Total Environ 761:143216. ISSN 0048-9697. https://doi.org/10.1016/j.scitotenv.2020.143216. https://www.sciencedirect.com/science/article/pii/S0048969720367474

Igual L, Segui S (2024) Introduction to data science. In: Introduction to data science: a python approach to concepts, techniques and applications, pp 1–4. Springer

Jazrawi C, Biller P, He Y, Montoya A, Ross AB, Maschmeyer T, Haynes BS (2015) Two-stage hydrothermal liquefaction of a high-protein microalga. Algal Res 8:15–22. ISSN 2211-9264. https://doi.org/10.1016/j.algal.2014.12.010. https://www.sciencedirect.com/science/article/pii/S2211926414001428

Katongtung T, Onsree T, Tippayawong N (2022) Machine learning prediction of biocrude yields and higher heating values from hydrothermal liquefaction of wet biomass and wastes. Bioresour Technol 344:126278. ISSN 0960-8524. https://doi.org/10.1016/j.biortech.2021.126278. https://www.sciencedirect.com/science/article/pii/S0960852421016205

Khuenkaeo N, MacQueen B, Onsree T, Daiya S, Tippayawong N, Lauterbach J (2020) Bio-oils from vacuum ablative pyrolysis of torrefied tobacco residues. RSC Adv 10:34986–34995. DOI:https://doi.org/10.1039/D0RA06014C. http://dx.doi.org/10.1039/D0RA06014C

Kruse A, Maniam P, Spieler F (2007) Influence of proteins on the hydrothermal gasification and liquefaction of biomass. 2. model compounds. Ind Eng Chem Res 46(1):87–96. ISSN 0888-5885. https://doi.org/10.1021/ie061047h. https://doi.org/10.1021/ie061047h

Li J, Pan L, Suvarna M, Tong YW, Wang X (2020) Fuel properties of hydrochar and pyrochar: Prediction and exploration with machine learning. Appl Energy 269:115166. ISSN 0306-2619. https://doi.org/10.1016/j.apenergy.2020.115166. https://www.sciencedirect.com/science/article/pii/S0306261920306784

Li J, Zhang W, Liu T, Yang L, Li H, Peng H, Jiang S, Wang X, Leng L (2021) Machine learning aided bio-oil production with high energy recovery and low nitrogen content from hydrothermal liquefaction of biomass with experiment verification. Chem Eng J 425:130649. ISSN 1385-8947. https://doi.org/10.1016/j.cej.2021.130649. https://www.sciencedirect.com/science/article/pii/S138589472102235X

McCullough K, Williams T, Mingle K, Jamshidi P, Lauterbach J (2020) High-throughput experimentation meets artificial intelligence: a new pathway to catalyst discovery. Phys Chem Chem Phys 22:11174–11196. DOI:https://doi.org/10.1039/D0CP00972E. http://dx.doi.org/10.1039/D0CP00972E

Mehrabadi A, Craggs R, Farid MM (2017) Wastewater treatment high rate algal pond biomass for bio-crude oil production. Bioresour Technol 224:255–264. ISSN 0960-8524. https://doi.org/10.1016/j.biortech.2016.10.082. https://www.sciencedirect.com/science/article/pii/S0960852416314961

Mishra S, Mohanty K (2020) Co-htl of domestic sewage sludge and wastewater treatment derived microalgal biomass - an integrated biorefinery approach for sustainable biocrude production. Energy Convers Manage 204:112312. ISSN 0196-8904. https://doi.org/10.1016/j.enconman.2019.112312. https://www.sciencedirect.com/science/article/pii/S0196890419313184

Neveux N, Yuen A, Jazrawi C, Magnusson M, Haynes B, Masters A, Montoya A, Paul N, Maschmeyer T, de Nys R (2014) Biocrude yield and productivity from the hydrothermal liquefaction of marine and freshwater green macroalgae. Bioresour Technol 155:334–341. ISSN 0960-8524. https://doi.org/10.1016/j.biortech.2013.12.083. https://www.sciencedirect.com/science/article/pii/S0960852413019238

Onsree T, Tippayawong N (2020) Torrefaction of maize residue pellets with dry flue gas. BioEnergy Research 13(1):358–368. ISSN 1939-1242. https://doi.org/10.1007/s12155-019-10058-x. https://doi.org/10.1007/s12155-019-10058-x

Onsree T, Tippayawong N (2021) Machine learning application to predict yields of solid products from biomass torrefaction. Renewable Energy 167:425–432. ISSN 0960-1481. https://doi.org/10.1016/j.renene.2020.11.099. https://www.sciencedirect.com/science/article/pii/S0960148120318516

Onsree T, Sittisun P, Sasaki R, Tippayawong N (2018a) Pyrolysis of corn residues: Kinetic analysis using discrete distributed activation energy model. IOP Conf Ser Earth Environ Sci 159(1):012036. https://doi.org/10.1088/1755-1315/159/1/012036.https://dx.doi.org/10.1088/1755-1315/159/1/012036

Onsree T, Tippayawong N, Zheng A, Li H (2018b) Pyrolysis behavior and kinetics of corn residue pellets and eucalyptus wood chips in a macro thermogravimetric analyzer. Case Studies in Thermal Engineering 12:546–556. ISSN 2214-157X. https://doi.org/10.1016/j.csite.2018.07.011. https://www.sciencedirect.com/science/article/pii/S2214157X18301424

Onsree T, Tippayawong N, Williams T, McCullough K, Barrow E, Pogaku R, Lauterbach J (2019) Torrefaction of pelletized corn residues with wet flue gas. Bioresour Technol 285:121330. ISSN 0960-8524. https://doi.org/10.1016/j.biortech.2019.121330. https://www.sciencedirect.com/science/article/pii/S0960852419305528

Onsree T, Jaroenkhasemmeesuk C, Tippayawong N (2020) Techno-economic assessment of a biomass torrefaction plant for pelletized agro-residues with flue gas as a main heat source. Energy Rep 6:92–96. ISSN 2352-4847. https://doi.org/10.1016/j.egyr.2020.10.043. https://www.sciencedirect.com/science/article/pii/S2352484720313962. 2020 The 7th International Conference on Power and Energy Systems Engineering

Pathy A, Meher S, Balasubramanian P (2020) Predicting algal biochar yield using extreme gradient boosting (xgb) algorithm of machine learning methods. Algal Res 50:102006. ISSN 2211-9264.

https://doi.org/10.1016/j.algal.2020.102006. https://www.sciencedirect.com/science/article/pii/S2211926420303684

Pedregosa F, Varoquaux G, Gramfort A, Michel V, Thirion B, Grisel O, Blondel M, Prettenhofer P, Weiss R, Dubourg V, Vanderplas J, Passos A, Cournapeau D, Brucher M, Perrot M, Duchesnay E (2011) Scikit-learn: machine learning in python. J Mach Learn 12:2825–2830

Phromphithak S, Onsree T, Tippayawong N (2021) Machine learning prediction of cellulose-rich materials from biomass pretreatment with ionic liquid solvents. Bioresour Technol 323:124642. ISSN 0960-8524. https://doi.org/10.1016/j.biortech.2020.124642. https://www.sciencedirect.com/science/article/pii/S0960852420319167

Ponnusamy VK, Nagappan S, Bhosale RR, Lay C-H, Duc Nguyen D, Pugazhendhi A, Chang SW, Kumar G (2020) Review on sustainable production of biochar through hydrothermal liquefaction: Physico-chemical properties and applications. Bioresour Technol 310:123414. ISSN 0960-8524. https://doi.org/10.1016/j.biortech.2020.123414. https://www.sciencedirect.com/science/article/pii/S0960852420306866

Rajagopal J, Gopinath KP, Krishnan A, Vikas Madhav N, Arun J (2021) Photocatalytic reforming of aqueous phase obtained from liquefaction of household mixed waste biomass for renewable bio-hydrogen production. Bioresour Technol 321:124529. ISSN 0960-8524. https://doi.org/10.1016/j.biortech.2020.124529. https://www.sciencedirect.com/science/article/pii/S0960852420318034

Reddy HK, Muppaneni T, Ponnusamy S, Sudasinghe N, Pegallapati A, Selvaratnam T, Seger M, Dungan B, Nirmalakhandan N, Schaub T, Holguin FO, Lammers P, Voorhies W, Deng S (2016) Temperature effect on hydrothermal liquefaction of nannochloropsis gaditana and chlorella sp. Appl Energy 165:943–951. ISSN 0306-2619. https://doi.org/10.1016/j.apenergy.2015.11.067. https://www.sciencedirect.com/science/article/pii/S0306261915015172

Saengsuriwong R, Onsree T, Phromphithak S, Tippayawong N (2021) Biocrude oil production via hydrothermal liquefaction of food waste in a simplified high-throughput reactor. Biores Technol 341:125750. ISSN 0960-8524. https://doi.org/10.1016/j.biortech.2021.125750. https://www.sciencedirect.com/science/article/pii/S0960852421010919

Saengsuriwong R, Onsree T, Phromphithak S, Tippayawong N (2023) Conversion of tobacco processing waste to biocrude oil via hydrothermal liquefaction in a multiple batch reactor. Clean Technol Environ Policy 25(2):397–407. ISSN 1618-9558. https://doi.org/10.1007/s10098-021-02132-w. https://doi.org/10.1007/s10098-021-02132-w

Tang X, Zhang C, Yang X (2020) Optimizing process of hydrothermal liquefaction of microalgae via flash heating and isolating aqueous extract from bio-crude. J Clean Prod 258:120660. ISSN 0959-6526. https://doi.org/10.1016/j.jclepro.2020.120660. https://www.sciencedirect.com/science/article/pii/S0959652620307071

Tanyaket T, Onsree T, Tippayawong N, Baratieri M (2020) Effect of oxidative torrefaction on characteristics of treated corncob pellets. J Chin Soc Mech Eng Trans Chin Inst Eng 41(1):65–73

Tian C, Liu Z, Zhang Y, Li B, Cao W, Lu H, Duan N, Zhang L, Zhang T (2015) Hydrothermal liquefaction of harvested high-ash low-lipid algal biomass from dianchi lake: Effects of operational parameters and relations of products. Biores Technol 184:336–343. ISSN 0960-8524. https://doi.org/10.1016/j.biortech.2014.10.093. https://www.sciencedirect.com/science/article/pii/S0960852414015259. Advances in biofuels and chemicals from algae

Tippayawong N, Onsree T, Williams T, McCullough K, MacQueen B, Lauterbach J (2021) Catalytic torrefaction of pelletized agro-residues with cu/al2o3 catalysts. Biomass Conversion and Biorefinery 11(5):1847–1852. ISSN 2190-6823. https://doi.org/10.1007/s13399-019-00535-w. https://doi.org/10.1007/s13399-019-00535-w

Torri C, Garcia Alba L, Samori C, Fabbri D, Brilman DWFW (2012) Hydrothermal treatment (htt) of microalgae: Detailed molecular characterization of htt oil in view of htt mechanism elucidation. Energy Fuels 26(1):658–671. ISSN 0887-0624. https://doi.org/10.1021/ef201417e. https://doi.org/10.1021/ef201417e

Yang J, He QS, Niu H, Corscadden K, Astatkie T (2018) Hydrothermal liquefaction of biomass model components for product yield prediction and reaction pathways exploration. Appl Energy

228:1618–1628. ISSN 0306-2619. https://doi.org/10.1016/j.apenergy.2018.06.142. https://www.sciencedirect.com/science/article/pii/S0306261918310110
Yang W, Wang Z, Han J, Song S, Zhang Y, Gong W (2019) The role of polysaccharides and proteins in bio-oil production during the hydrothermal liquefaction of algae species. RSC Adv 9:41962–41969. DOI:https://doi.org/10.1039/C9RA07150D.http://dx.doi.org/10.1039/C9RA07150D
Yeo I-K, Johnson RA (2000) A new family of power transformations to improve normality or symmetry. Biometrika 87(4):954–959. ISSN 00063444. http://www.jstor.org/stable/2673623
Zhu X, Li Y, Wang X (2019) Machine learning prediction of biochar yield and carbon contents in biochar based on biomass characteristics and pyrolysis conditions. Biores Technol 288:121527. ISSN 0960-8524. https://doi.org/10.1016/j.biortech.2019.121527. https://www.sciencedirect.com/science/article/pii/S0960852419307576

Index

N. Tippayawong et al., *Machine Learning Applications for Bioenergy Conversion*, Green Energy and Technology, https://doi.org/10.1007/978-981-96-8086-3

MIX
Papier aus verantwortungsvollen Quellen
Paper from responsible sources
FSC® C105338

If you have any concerns about our products,
you can contact us on
ProductSafety@springernature.com

In case Publisher is established outside the EU,
the EU authorized representative is:
Springer Nature Customer Service Center GmbH
Europaplatz 3, 69115 Heidelberg, Germany

Printed by Libri Plureos GmbH
in Hamburg, Germany